Geology of the Peterborough district

The district described in the memoir lies at the western margin of the Fenland. Most of the countryside is devoted to arable farming. Originally an administrative centre and market town, the expanding city of Peterborough is an increasingly important centre of manufacturing industry and warehouse facilities.

The Fenland is underlain by Flandrian deposits which generally lie below 2 m OD but occur in places below Ordnance Datum. Within the low-lying area, there are slightly raised 'islands' which consist of drift deposits capping low rises formed of Upper Jurassic clays. The 'uplands' in the south-west, which embrace the area of urban Peterborough, consist of Middle Jurassic rocks.

There are very few natural exposures and much of the new information has been derived from boreholes for the site investigation industry and the tapping of the water resources of the Lincolnshire Limestone.

Precambrian or Cambrian volcanic and metamorphosed quartzitic sediments occur at depth. These are overlain by Triassic continental and marine sediments. After the main Rhaetian marine transgression there was a major retreat until, during Liassic times, the sea again migrated south-eastwards, progressively inundating the London Platform. During Middle Jurassic times the environment varied from restricted marine to deltaic and freshwater. More uniform marine conditions were re-established in Upper Jurassic times.

There was a long hiatus until the Pleistocene glaciers entered the Fenland during the Anglian stage and deposited sediments of Chalky Boulder Clay type. A period of downcutting followed with the establishment of the modern drainage system and the accumulation of river terrace deposits and their equivalent marine gravels. The final stage was the accumulation of peats with, nearer the sea, salt marsh and sand bar and creek sediments of the Barroway Drove Beds and the Terrington Beds.

BRITISH GEOLOGICAL SURVEY

A. HORTON

Geology of the Peterborough district

CONTRIBUTOR
R. A. Downing

Memoir for 1:50 000 geological sheet 158 (England and Wales)

LONDON: HER MAJESTY'S STATIONERY OFFICE 1989

First published 1989

ISBN 0 11 884409 1

Bibliographical reference

HORTON A. 1989. Geology of the Peterborough district. *Mem. Br. Geol. Surv.*, Sheet 158, England and Wales.

Authors

A. HORTON, BSc
British Geological Survey, Keyworth

Contributor

R. A. Downing, DSc
formerly British Geological Survey, Wallingford

Other publications of the Survey dealing with this district and adjoining districts

BOOKS

British Regional Geology
Eastern England from the Tees to the Wash (Edition2)
Central England (Edition 3)
East Anglia and adjoining areas (Edition 4)

Sheet Memoirs
Geology of the Fenland
Geology of the country around Kettering, Sheet 171

MAPS

1:625 000
Solid Geology (South sheet)
Quaternary geology (South sheet)
Aeromagnetic map (South sheet)
Hydrogeologic map (2, North and East Lincolnshire)

1:250 000
East Midlands (Aeromagnetic)
East Midlands (Gravity)
East Midlands (Solid Geology)

1:50 000 (and one inch to one mile)
Sheet 143 Bourne (1964)
Sheet 145, part 129 King's Lynn and The Wash (1978)
Sheet 157 Stamford (1978)
Sheet 171 Kettering (1976)
Sheet 173 Ely (1980)

1:25 000
Peterborough Sheet Parts of TF 00, 10, 20 and TL 09, 19, 29 (1979)

Printed in the United Kingdom for Her Majesty's Stationery Office
Dd 240412 C20 3/89 12521

CONTENTS

FIGURES

TABLES

PLATES

NOTES

National Grid references given in the form [1502 0526] lie within the 100-km square TF. Grid references prefaced by the letters TL lie within that 100 km square.

PREFACE

The original geological survey of the Peterborough district was made at the scale of 1 inch to 1 mile by S. B. J. Skertchly, J. W. Judd and W. H. Holloway. The results were published on Old Series sheets 64 (eastern part) in 1872 and 65 (western part) in 1886. Skertchly described the geology of the Fenland in 1877.

The resurvey of the district at the scale of 6 inches to 1 mile commenced in 1945 with an investigation of the resources of Jurassic ironstone by Mr G. Bisson, Mr W. B. Evans, Professor J. H. Taylor, Mr R. F. Goossens and Dr P. A. Sabine. The resurvey was extended in 1968 by Mr A. Horton, Mr R. D. Lake and Dr B. C. Coppack over the area designated for the construction of the new town of Greater Peterborough. The work was completed during the period 1976–79 by Mr Horton, Dr E. R. Shephard-Thorn, Dr J. M. Ridgway, Mr C. J. Wilcox, Mr R. J. Wyatt and Dr J. A. Zalasiewicz with the mapping of the Fenlands. Subsequently Mr S. J. Booth undertook an assessment of the sand and gravel resources of the district and its environs on behalf of the Department of the Environment and the results were published in Mineral Assessment Reports in 1981, 1982 and 1983. The geological map was published in 1984 at the scale of 1:50 000. This Memoir was designed as a Sheet Explanation and is intended only as a brief guide to the geology of the district. It does not seek to provide the detailed descriptions of a conventional sheet memoir and the full field and other data are available for reference in the National Geosciences Data Centre at Keyworth, Nottinghamshire.

This Memoir has been written and compiled by Mr Horton from the notes of other surveyors; Dr R. A. Downing contributed the account of the water supply and diagrams have been drafted by Mr J. P. Colleran (Figure 1), Mr Booth (Figure 15A), Mr Wyatt (Figures 13 and 15B) and Dr Zalasiewicz (Figure 16). The Memoir has been edited by Mr J. E. Wright.

In addition, many individuals, companies and authorities, too numerous to mention here, have contributed information and advice for which we are most grateful. The work of the surveyors was also greatly eased by the generous assistance and co-operation of landowners and tenants in providing access to their properties.

F. G. Larminie, OBE
Director

British Geological Survey,
Keyworth, Nottinghamshire
NG12 5GG

1 November 1988

LIST OF WORKING QUARRIES

SAND AND GRAVEL

First Terrace Deposits

LANGTOFT COMMON [TF 1395 1475]. Amey Roadstone Corporation, Baston Pit, Peterborough

BASTON [TF 1430 1360]. F. B. Gibbons Ltd, Manor Pit, Baston, Peterborough

MAXEY [TF 1300 0760]. Tarmac Ltd, Barn Farm Pit, Maxey, Peterborough

EYE, Tanholt Farm [TF 2304 0145]. Roade Aggregates Ltd, Tanholt Farm Pit, Eyebury Road, Eye, Peterborough

THORNEY, Willow Hall [TF 2520 0090]. Amey Roadstone Corporation, Baston Pit, Peterborough

Second Terrace Deposits

WHITTLESEY area. Extraction, in stages, by contract with London Brick plc prior to enlargement of the Saxon and Kings Dyke Claypits—see below

BRICK CLAY

Oxford Clay

All pits worked by London Brick plc, Stewartby, Bedfordshire

DOGSTHORPE [TF 204 021]

SAXON [TL 245 975]

KINGS DYKE [TL 240 970]

Permission to visit the quarries must be obtained from the owners.

LIST OF SIX-INCH MAPS

Geological six-inch maps included wholly or in part in the 1:50 000 Sheet 158 (Peterborough) are listed below together with the initials of the surveyors and the dates of the survey. The surveyors were G. Bisson, B. C. Coppack, W. B. Evans, R. F. Goossens, A. Horton, R. D. Lake, J. M. Ridgway, P. A. Sabine, E. R. Shephard-Thorn, J. H. Taylor, C. J. Wilcox, V. Wilson, R. J. Wyatt and J. A. Zalasiewicz. Manuscript copies of these maps are available for public reference in the libraries of the British Geological Survey in Keyworth and Edinburgh.

Map	Surveyors	Dates
TF 10 NW	GB,WBE,JHT,RJW	1946–47, 1977
NE	GB,RJW	1947, 1977
SW	GB,JHT	1946–47
SE	GB,RDL	1947, 1968
TF 11 NW	WBE,VW,RJW	1946–54, 1977
NE	RJW	1977–78
SW	RJW	1977
SE	RJW	1977
TF 20 NW	JMR	1976
NE	JMR	1976
SW	RDL,JMR	1968, 1976
SE	JMR	1976
TF 21 NW	RJW	1978
NE	AH	1979
SW	CJW	1979
SE	CJW	1979
TF 30 NW	JAZ	1979
NE	JAZ	1979
SW	RJW	1979
SE	RJW	1979
TF 31 NW	AH	1979
NE	AH	1979
SW	AH	1979
SE	CJW,JAZ	1979
TF 40 NW	AH,JAZ	1979
SW	RJW	1979
TF 41 NW	AH	1979
SW	AH	1979
TL 19 NW	RFG,PAS,BCC	1945–47, 1968
NE	GB,AH	1947, 1968
TL 29 NW	AH, JMR	1968, 1976
NE	JMR	1976
TL 39 NW	ERST	1979
NE	ERST	1979
TL 49 NW	ERST	1979

ONE

Introduction and geological history

The Peterborough district can be divided into three physiographic regions: the low-lying Fens which cover the greater part of the north and east, the intermediate benches formed by the deposits of river and marine gravel which are best developed along the margins of the Fenlands, and thirdly the Jurassic 'uplands' in the south-west. The Fens are relatively flat and generally lie less than 3 m above OD with parts below sea level, although small gravel outcrops may form benches at up to 10 m above OD. In contrast the undulating uplands rise to over 40 m above OD at Castor Hanglands and contain woodland and arable areas. The district is drained mainly by two rivers, the Welland in the north and the Nene, which flows through the uplands in the south and which is canalised downstream from its entry into the Fens at Peterborough.

The succession of rocks in the district is shown on the inside front cover and in the vertical section on the geological map. The solid rocks range in age from ?Precambrian to Upper Jurassic and are overlain by a sequence of drift deposits of Pleistocene and Holocene age. Figure 1 is a sketch map of the geology of the district.

Basement rocks were proved in the Glinton, Spalding No. 1 and Wisbech No. 1 boreholes. Volcanic rocks of possible Precambrian or Cambrian age were proved at Glinton and quartzites of similar age were penetrated at Spalding and Wisbech.

In Triassic times these ancient rocks were buried beneath a sequence of water-laid sandstones, conglomerates and mudstones, and at the close of the Triassic period the Rhaetian marine transgression covered the region. After a short period of regression, marine conditions were re-established and continued throughout the period of deposition of the Lias. During Middle Jurassic times there was a relative fall in sea level and shallow or restricted marine conditions were interspersed with periods of land emergence. Marine conditions were re-established during the late Middle Jurassic

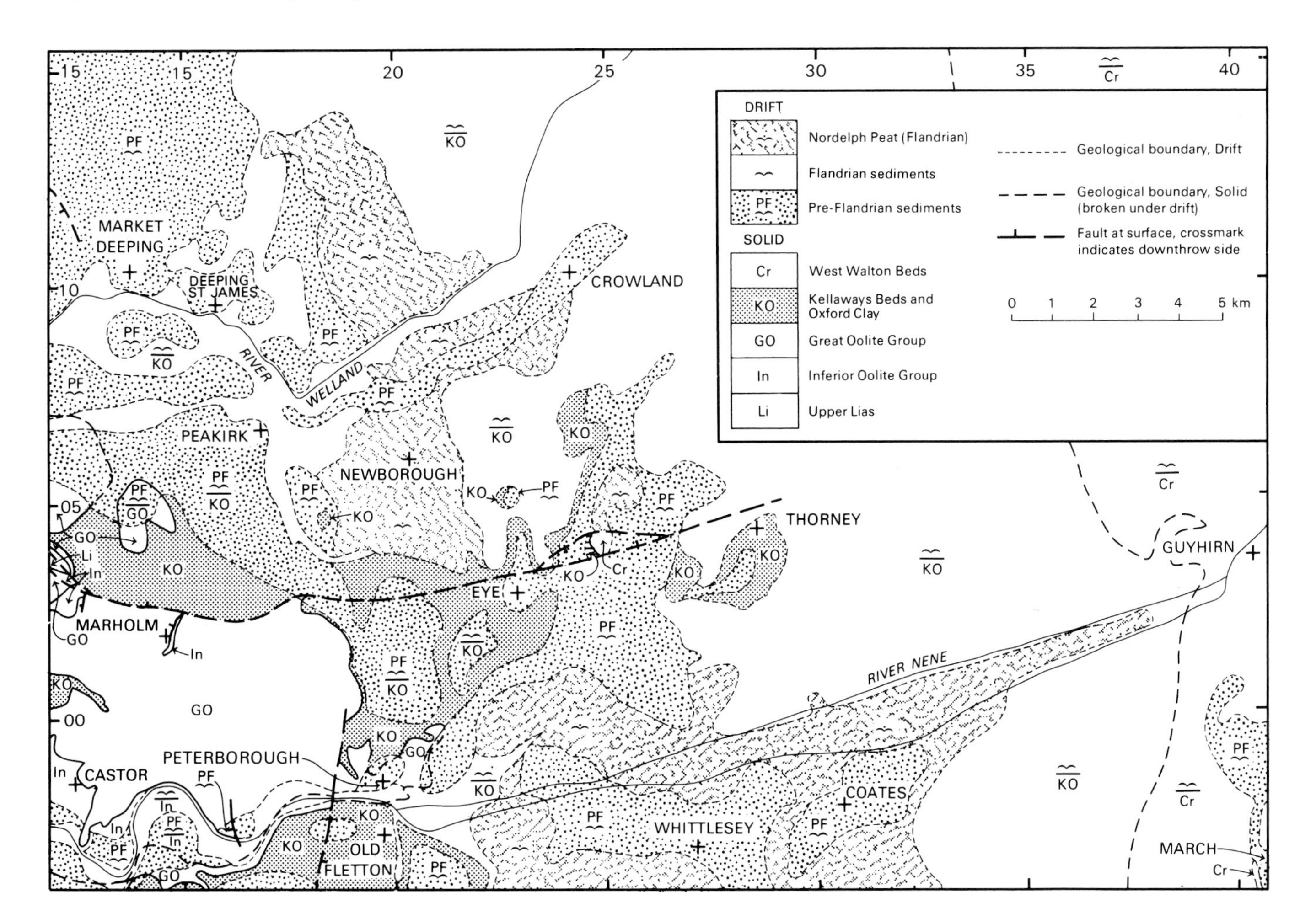

Figure 1 Sketch map of the geology of the district

and the depth of water continued to fluctuate throughout the Upper Jurassic. The Cornbrash and the Kellaways Sands were deposited in shallow, current-swept environments but, in contrast, the Kellaways and Oxford clays accumulated in deeper off-shore conditions. The West Walton Beds represent a slightly restricted and shallower phase of deposition.

The Jurassic strata were gently folded and faulted during the subsequent period up to the end of the Tertiary. Most of the folds appear to be associated with the faulting and are superimposed on a very gentle regional dip to the south-east.

The oldest Quaternary deposits were left by the glaciation which laid down the Chalky Boulder Clay. The ice sheet advanced across a landscape with a drainage system dating from late Tertiary to early Quaternary times, and covered the surface with glacially-derived debris. After the retreat of the ice, the present drainage was developed and, with progressive deposition of river gravel and subsequent down-cutting, the present pattern of river terraces was established. During the Flandrian, dating from about 10 000 years ago, a shallow sea advanced across the Fenlands and a sequence of estuarine silts and clays was deposited. These sediments interdigitate with peats and river alluvium in the areas nearer to the Fen margins.

TWO

Pre-Mesozoic rocks

Three boreholes proved pre-Mesozoic rocks within the Peterborough district. The Glinton Borehole [1502 0526] (Kent, 1962; Horton and others, 1974) proved flow-structured rhyolitic lava containing pyroclasts of crystal-tuff. Thin sections show magmatically corroded pyroclasts which were rotated and fractured during flow before being enclosed in a matrix of lava and tuff which has since been devitrified. Shards are not conspicuous and there is little evidence of melting although this may be obscured by the alteration products. Potassium-argon dating of two specimens gave ages of 386 ± 7 and 377 ± 10 million years. Agglomerates and tuffs of similar type have been proved in boreholes at Upwood, south-east of Peterborough, at North Creake, Norfolk (Kent, 1947) and near Stamford. All these rocks are similar to the Precambrian porphyroids of Charnwood Forest but the potassium-argon age of the Glinton specimens indicates that they may have been subject to mineralogical change during the Caledonian orogeny. Alternatively they may come from intrusions of Ordovician or Silurian age.

The Spalding No. 1 [2434 1478] and Wisbech No. 1 [4066 0842] boreholes proved pale grey quartzites whose age is uncertain. They could be of Cambrian or possibly Precambrian age.

Regional gravity surveys in the eastern Midlands indicate the presence of several negative Bouguer anomalies extending ESE from Nottingham to north Norfolk. Those extending north-eastwards from Peterborough are attributed to the presence of relatively light acid igneous bodies. Aeromagnetic surveys over the region reveal the existence of strong magnetic anomalies, the pattern of which suggests that the basement at depth may contain bodies of igneous rock with cupolas extending upwards to shallower depths. The Glinton volcanic rocks may be part of this igneous complex. Figure 2 is based on the results of gravity, aeromagnetic and seismic surveys, calibrated by the findings of boreholes which have penetrated beneath the Mesozoic (Smith and others, 1985). It shows the conjectural sub-Mesozoic outcrops and generalised contours in metres below OD on the base of Mesozoic.

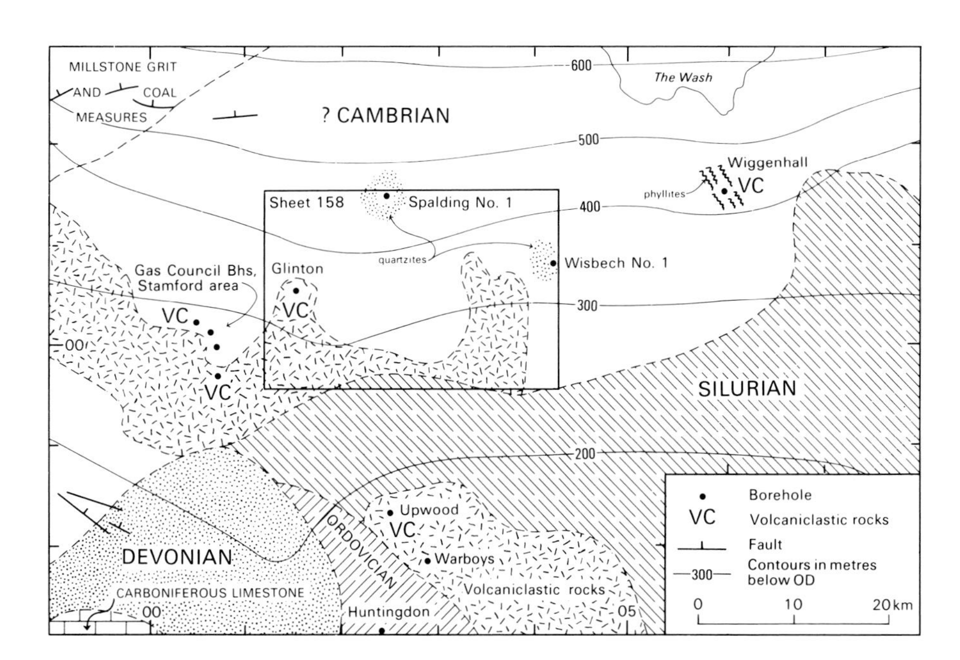

Figure 2 Pre-Mesozoic basement of the region around Peterborough

THREE

Triassic rocks

Triassic rocks do not crop out within the Peterborough district but are known from the Spalding No. 1 [2434 1478], Glinton [1502 0526] and Wisbech No. 1 [4066 0842] boreholes. Figure 3 illustrates the sequences found in these holes. It is thought that the Triassic is present at depth throughout the district, decreasing in thickness from over 120 m in the north-west to less than 20 m in the south. The Triassic deposits probably rest with strong unconformity upon the older rocks of the basement. The period was one of fluvial and lacustrine sedimentation the deposits of which gradually buried the earlier landscape.

The basal sediments are probably mainly sandstones and conglomerates which have been correlated tentatively with the Bromsgrove Sandstone Formation. In the Spalding Borehole about 41 m of pebbly sandstone passes down into a well cemented conglomerate. At Glinton about 6.4 m of fine-grained sandstone overlie about 9.5 m of conglomeratic sandstone, but in the Wisbech Borehole the basal member comprised about 15 m of fine-grained dolomitic sandstone.

The Mercia Mudstone Group, represented by reddish brown siltstones with fine-grained sandstones, was identified in the Spalding (59.7 m) and Glinton (16.7 m) boreholes. It was not proved in the Wisbech Borehole. The Penarth Group comprises the Westbury and Lilstock formations, parts of which have been identified in all three boreholes. The Westbury Formation comprises dark grey or black mudstones with thin, fine-grained sandstones and was estimated as 13.8 m in the Glinton Borehole and 7.3 m in the Spalding Borehole. It may also be represented within the 1.5 m of mudstones attributed to the Penarth Group in the Wisbech Borehole. The Westbury Formation was deposited in a marine environment which terminated the continental sedimentation that had characterised most of the Triassic Period. The overlying Cotham Member of the Lilstock Formation shows evidence of shallow inshore deposition, locally perhaps even in fresh water. It comprises greenish mudstones with silty partings and was found in the Spalding (3 m) and Glinton (9.5 m) boreholes. The Langport Member of the Lilstock Formation was identified only in the Wisbech Borehole as 2.1 m of grey bioclastic limestone but may also be present in the north-western margins of the district.

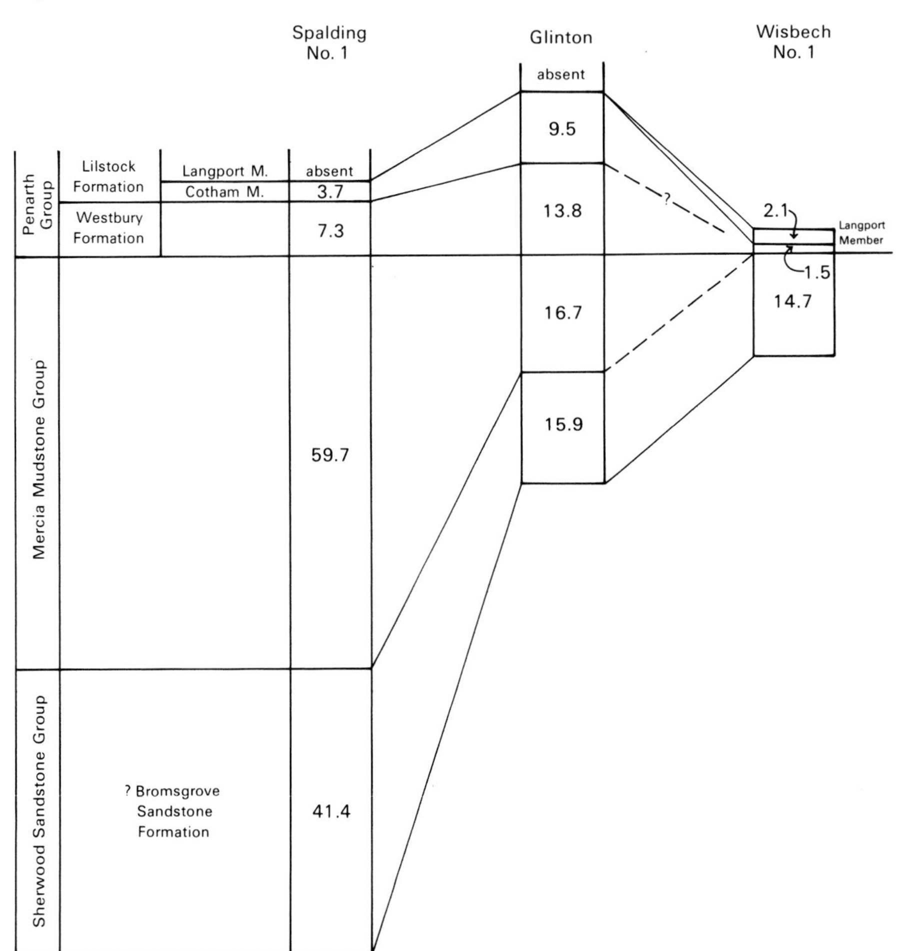

Figure 3 Triassic sequence in the Spalding, Glinton and Wisbech boreholes

FOUR

Jurassic rocks

Jurassic rocks occur throughout the Peterborough district but crop out mainly in the south-west corner. Elsewhere they are generally covered by Quaternary terrace deposits or by the peat and alluvium of the Fens. The generalised stratigraphical sequence is illustrated on the inside front cover where the major lithological groups are equated to the biostratigraphical stages, the latter being based mainly on ammonite zonation (Cope and others, 1980a and b). The total thickness of the Jurassic within the district ranges from about 400 m in the north-west to about 240 m in the south-east. This may indicate a contemporaneous decrease in the amount of deposition towards the London Platform, south-east of the district.

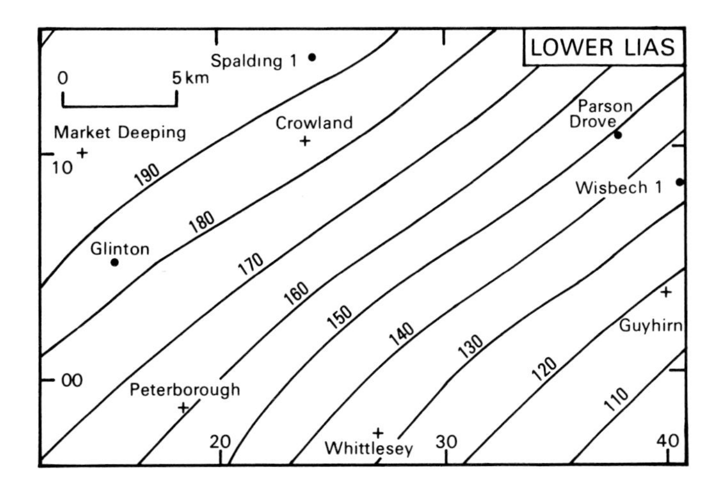

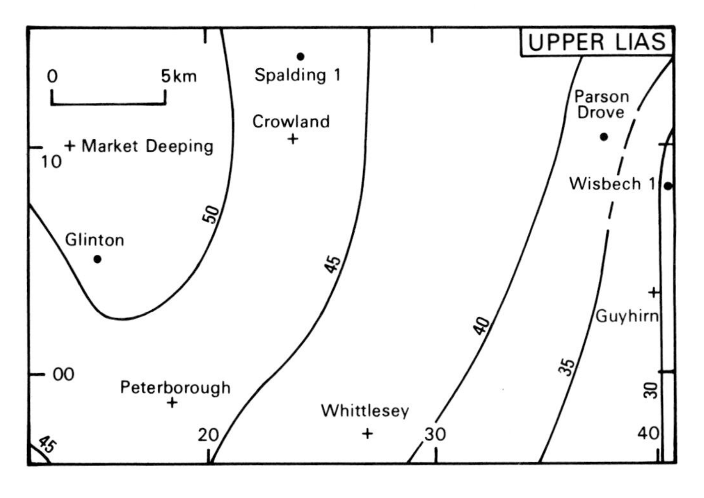

Figure 4 Thickness variations in the Lower and Upper Lias

LIAS GROUP

The Lias Group comprises lower, middle and upper divisions, but only the Upper Lias crops out within the district, south of Helpston on the western margin. The rest of the Group is known only from boreholes, almost all of which were not cored but were classified on the basis of drill cuttings and geophysical well logs. Figure 4 shows the variation in thickness of the Lower and Upper Lias. Figure 5 is a correlation of geophysical markers in the Lias Group from boreholes in the district.

Lower Lias

The Lower Lias is thickest in the north-west and thins to the south-east (Figure 4). A thickness of 186.2 m was recorded in the Glinton Borehole [1502 0526]. This differs from the figure of 156.4 m noted by Kent (1962, p.40) but his division is now considered to exclude the uppermost beds of Lower Lias. In the Spalding [2434 1478] and Wisbech [4066 0842] boreholes the estimated thicknesses were 191 and 136 m respectively.

The Lower Lias consists largely of grey mudstones with a variable carbonate content. Thin limestones in the lower part of the formation form persistent markers on gamma and resistivity logs and have been used for correlation in addition to higher lithologically distinctive horizons (Figure 5). The limestones are generally fine-grained but there are some horizons rich in shell debris. The Glinton, Spalding and Wisbech boreholes all proved limestones in the basal part of the Lias. Figure 5 correlates the limestones of Spalding and Glinton with the Hydraulic Limestones (of Nottinghamshire) but those of Wisbech may occur at a higher stratigraphical level. This site is nearer to the London Platform and may have been inundated by the Liassic marine transgression later than areas to the north-west.

The Lower Lias is considered to be entirely marine and to have accumulated in a mainly quiescent offshore environment.

Middle Lias

There is an upward gradation from the silty mudstones of the Lower Lias into the slightly arenaceous Middle Lias. The latter comprises a lower division of grey sandy silts and mudstones with sandy and silty limestones, the Middle Lias Silts and Clays, overlain by the Marlstone Rock Bed. The beds of the lower division are commonly laminated and bioturbated. The nearest cored borehole through the sequence was at Tydd St Mary [4307 1737], about 3 km north-east of the district. The Silts and Clays were estimated there to be 5.43 m thick, with a hard basal chamositic and sideritic bed which contained chamosite ooliths and phosphatic pebbles and which rested non-sequentially on the Lower Lias.

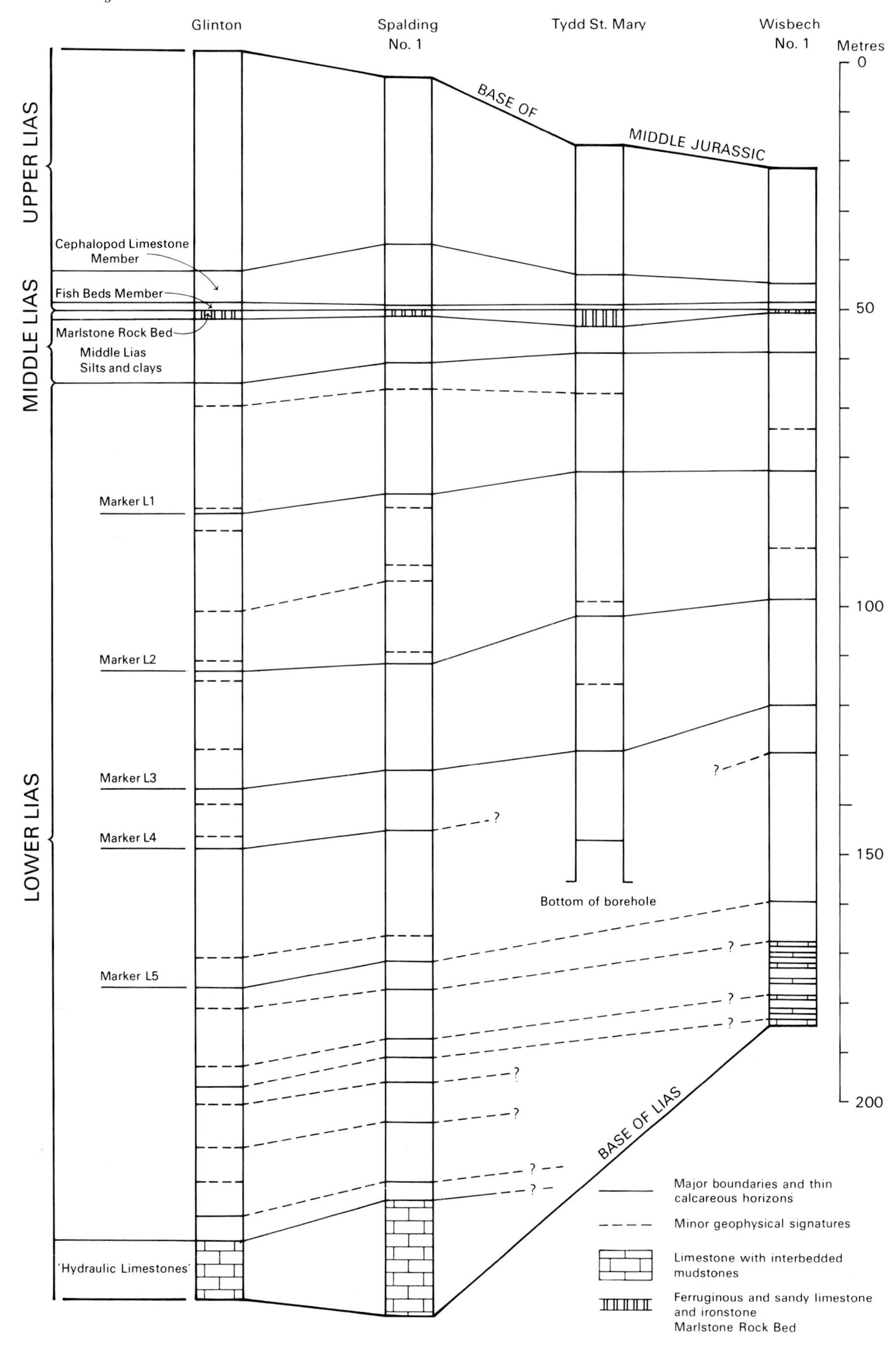

Figure 5 Correlation of geophysical markers in the Lias Group

Estimates of thickness in the uncored boreholes at Glinton, Spalding and Wisbech were about 13.7, 10.4 and 8.2 m respectively.

The Marlstone Rock Bed is a shelly limestone, sandy in part and locally with chamosite ooliths. It is estimated at up to 3 m thick and probably persists throughout the district but is not sufficently ferruginous to be an iron ore.

Marine conditions persisted throughout the period of deposition of the Middle Lias but were probably shallower than previously and associated with an influx of coarse silt and fine sand. A period of reworking of sediment was followed by limestone formation when chamositic ooliths were formed, or perhaps swept in from a more restricted marginal environment where silicate precipitation occurred.

Upper Lias

The Upper Lias consists mainly of grey mudstones, locally with thin limestones and occasional phosphatic and sideritic nodules. Only the topmost beds of the formation, mudstones with calcareous nodules and a thin lenticular limestones, crop out in the district, to the south of Helpston on the west margin of the district [120 036]. The basal Fish Beds Member is a dark grey bituminous shale with thin limestone bands and is overlain by ammonite-bearing, fine-grained limestones, the Cephalopod Limestones Member. Both members were recognised from cuttings in the Glinton Borehole and were also indicated by the geophysical records of the Spalding and Wisbech holes. In the cored Tydd St Mary Borehole the Upper Lias was 32.7 m thick and was overlain non-sequentially by the Lincolnshire Limestone.

The maximum thickness was estimated as 42.3 m in the Glinton Borehole and the formation thins south-eastwards to about 30 m (Figure 4). In the Parson Drove Borehole [3793 1052] only 6 m of Upper Lias were cored, mainly of the basal Fish Beds Member. The absence of higher beds here is attributed to erosion prior to the deposition of the overlying Lincolnshire Limestone.

The bituminous Fish Beds Member is interpreted as marking a return to quiescent depositional conditions, probably almost anaerobic. Conditions later became more varied during the deposition of the Cephalopod Limestones Member but probably became mainly quiescent subsequently in an open sea in which bottom-dwelling faunas were rare.

The youngest beds of the Upper Lias in the East Midlands belong to the *Hildoceras bifrons* Zone of early-Toarcian age.

INFERIOR OOLITE GROUP

The Inferior Oolite Group comprises the Northampton Sand, Grantham and Lincolnshire Limestone formations and ranges in age from Aalenian to mid-Bajocian. There is a local stratigraphical hiatus between the mid-Toarcian Upper Lias and the base of the Northampton Sand, due to erosion and possibly non-deposition. The stable marine environment of the Upper Lias was succeeded in Inferior Oolite times by shallower conditions with a fluctuating sea level, giving rise to littoral, deltaic or even fresh water sediments.

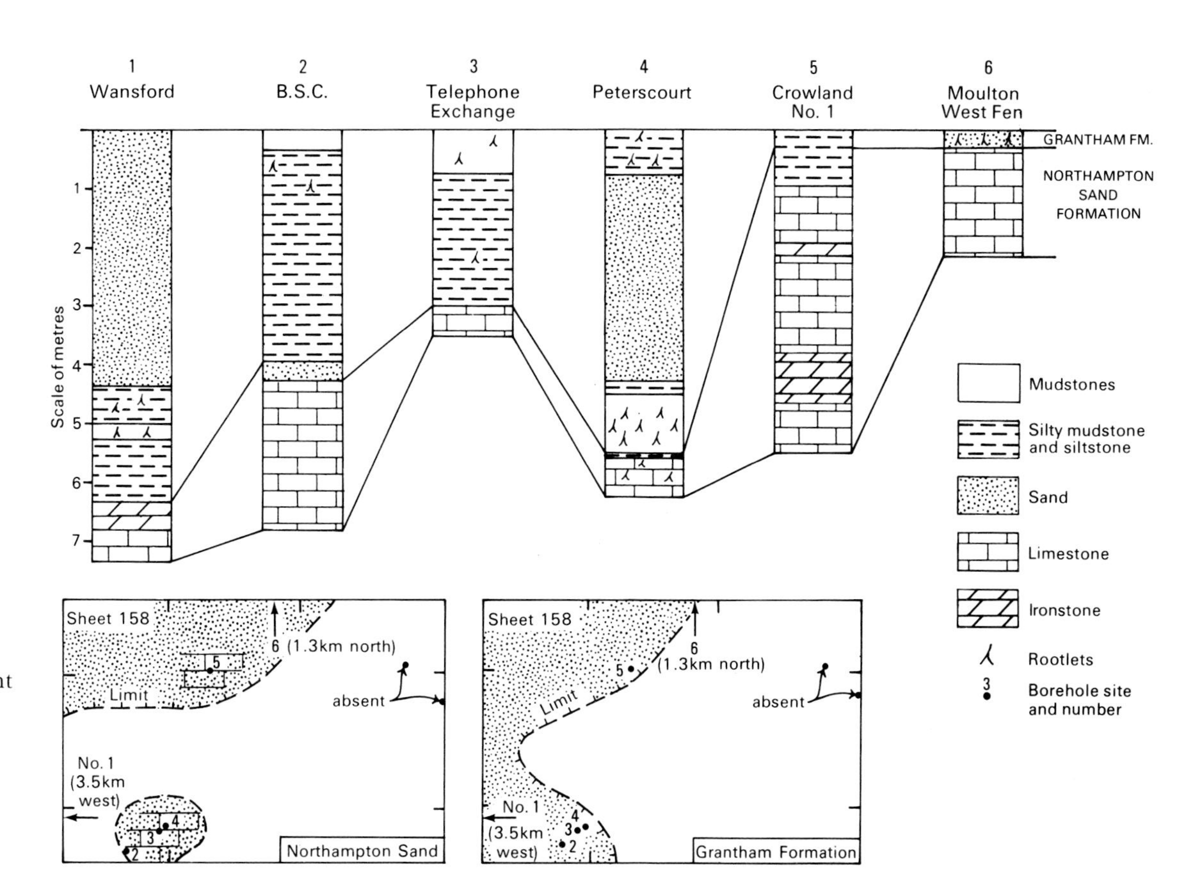

Figure 6 Extent and lithological variation of the Northampton Sand and Grantham Formation

Northampton Sand

The Northampton Sand does not crop out within the Peterborough district but is present at depth in the north-west part and probably as an isolated subcrop beneath south-west Peterborough (Figure 6). The sequence is very variable but includes silty and sandy limestones, locally shelly and oolitic, with some horizons of silty mudstone containing phosphatic nodules, cementstone pebbles and chamosite (berthierine) ooliths. Extensive bioturbation is common and there is plant debris at some levels while at others there are seatearths with abundant rootlets. These latter tend to be associated with the development of sphaerosiderite.

The local maximum thickness of 5.2 m of silty and sandy limestone was proved in Crowland No. 1 Borehole [2298 1067]. To the west, at Elm Farm, north of Deeping Gate, a borehole [1521 1290] proved 4.47 m of similar beds without reaching the base. Boreholes in the south-west Peterborough area (Figure 6) proved from about 0.6 m to about 2.8 m of variable sandy limestone and sand, with some clayey and sideritic horizons. These beds are bioturbated and contain plant debris and rootlets.

To the west of the district the Northampton Sand has been exploited for iron ore. There it comprises chamositic (berthierinitic) and sideritic oolitic limestones and mudstones which pass laterally into ferruginous sands. In the Peterborough district the sediments are considered to represent a marginal facies. The presence of rootlets penetrating what are now hard sandy limestones, and the secondary segregation of siderite around rootlet traces, indicate that the plants grew before completion of the diagenesis of the calcareous sediments. The presence of a non-sequence within these marginal deposits is indicated by the truncation of rootlets and the boring of the underlying layers in a borehole at Peterborough Telephone Exchange [TL 1905 9850]. The ironstone facies was deposited in a restricted environment where iron silicates were precipitated, but this gave way, to the east and south-east, to shallow marginal conditions where the sandy limestone facies of the present district was laid down in waters which were locally sufficiently shallow to allow colonisation by plants.

Grantham Formation

The Grantham Formation (formerly the Lower Estuarine 'Series') crops out [120 037] south of Helpston and at Castor [TL 127 982] west of Peterborough. It underlies the western part of the district from north of Crowland to south-west of Peterborough (Figure 6). The eastward limit of the Formation shown in this figure differs from the horizontal sections on the geological map. In some places the formation overlaps the Northampton Sand to rest directly upon the Upper Lias. It is itself overlapped by the Lincolnshire Limestone or may locally be absent because of channelling beneath the latter. Where the Lincolnshire Limestone is absent the Grantham Formation may be overlain with slight unconformity by the Upper Estuarine 'Series'.

The Formation consists of fine sands, silts, silty clays and mudstones. The sandy beds are micaceous and black or dark grey, but may also be white or pale grey. The darker colours probably weather to paler hues at outcrop. The silts are grey or brownish grey and the clays vary from brownish grey to black. The finer beds may be well laminated but other horizons are poorly sorted and not well bedded. There are also contorted beds with detached, slumped or load-cast masses of sand. Lignitic plant detritus is ubiquitous, rootlets are common and there are beds with seatearth texture and abundant listric surfaces.

The base of the Formation is distinct where it overlies the ironstone facies of the Northampton Sand but is less clear against the rooty and sandy facies of the latter, which is found in the present district. There is also difficulty where it rests on Upper Lias mudstone which has been converted to seatearth lithology with sphaerosiderite. Where the Lincolnshire Limestone overlies the Grantham Formation, the junction is commonly abrupt but it may be less clear in some outcrops where a basal sandy facies of the Limestone, the Sproxton Member, overlies sandy beds of the Grantham Formation. Where the Lincolnshire Limestone is absent there may also be difficulty in defining the top of the Grantham Formation, if it and the overlying Upper Estuarine 'Series' both consist of non-marine siltstones and mudstones.

It is thought that the maximum thickness occurs in the Castor–Peterborough area and about 7.6 m sand was recorded in an old water borehole [TL 1326 9821] south-east of Castor. The topmost bed of the Formation was exposed in a trench in Castor [TL 1271 9826] where it consisted of a black and purple-speckled, silty clay.

Only plant fossils and debris have been recognised locally. Taylor (1963, p.42) concluded that the depositional environment was a low-lying coastal plain or delta flat on the north-west side of the London Platform. The Stainby Shale Member represents a marine incursion developed in the Formation to the north-west of the present district.

Lincolnshire Limestone

The Lincolnshire Limestone crops out on each side of the Marholm Fault, also called the Tinwell–Marholm Fault, south of Helpston on the western margin of the district, and farther south on the lower slopes of the Nene valley at Castor. The main development of the Formation lies north of the Peterborough Disrict and it occurs at depth in the northern and western parts of the district, being absent in the south-east (Figure 7).

The lower part of the Lincolnshire Limestone consists of fine-grained sandy limestones with minor shell debris. These are thinly bedded and contain partings with mica and plant debris. At outcrop these strata are decalcified to pale brown sand which may contain residual rounded calcareous masses (potlids). This is the basal Sproxton Member of the formation and it is equated with the Collyweston Slate facies which occurs to north-west of the district. The overlying limestones are oolitic, commonly with coarse and fine ooliths scattered throughout the rock, but generally with only small amounts of shell debris (Horton and others, 1974, p.13). Beds of this type were proved in most of the cored boreholes in the district and were equated with the Lower Lincolnshire Limestone. Pebbles occur at various horizons but are not necessarily associated with the erosional and hardground surfaces recorded in the cores.

In the Parson Drove Borehole [3793 1052] an unexpected

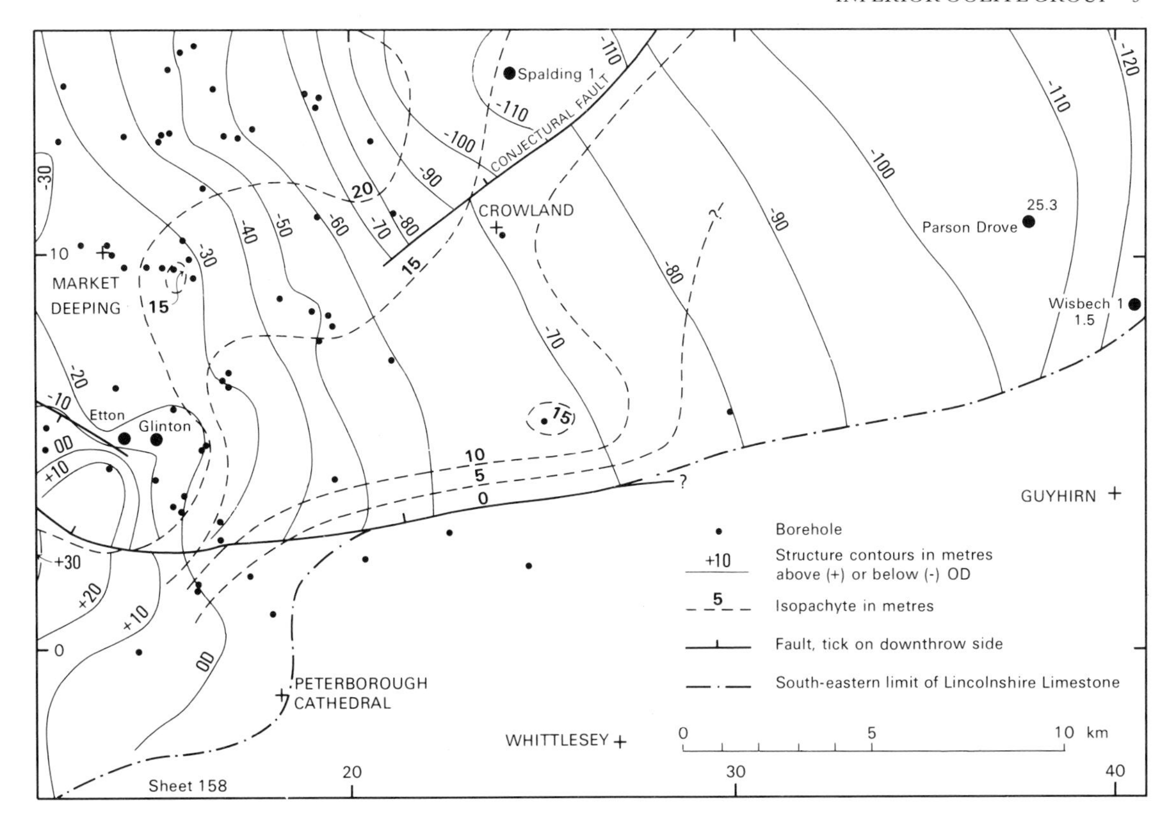

Figure 7 Thickness of the Lincolnshire Limestone and structure contours on the base

thickness of 25.35 m was proved, consisting of banded oolitic limestone with shell debris and scattered whole shells and ranging from friable poorly consolidated sediments to well cemented limestones. These beds probably occupy an erosional channel which was cut down below the surface of the Upper Lias. The absence of a basal Sproxton Member suggests a correlation with the Upper Lincolnshire Limestone.

Figure 7 shows that the thickness in the north-west part of the district exceeds 20 m. Boreholes near Etton Waterworks [1417 0511 and 1412 0497], south-west of Glinton, proved 24.5 m and at least 23.8 m respectively, although a partly cored borehole [1424 0518] indicated a thickness of 22 m hereabouts.

A hole at Marholm Crossing [1542 0356] found 23.4 m of the formation. To the north of Eye [230 030] there is a rapid reduction in thickness southwards from 10 m to a pinch-out close to the line of the Marholm Fault. It is possible that this structure may have influenced sedimentation in Lincolnshire Limestone times. Channels may be the cause of high thickness values in the Hop Pole area [180 140], as in the Parson Drove Borehole. The Lincolnshire Limestone generally overlies the Grantham Formation but it may overlap the latter to rest on the Northampton Sand or, in areas of erosional channels, the Upper Lias. Ammonite evidence from the Stamford area to the west indicates that the local Lincolnshire Limestone is of Lower Bajocian age.

Table 1 shows the division of the Lincolnshire Limestone according to Taylor (1963) from work in the Kettering area, and according to Ashton (1980) from work over a wider area. Taylor's divisions were used in the 1946 survey of the outcrop areas of the Peterborough District. Ashton's divisions are based on erosion surfaces which separate the members of the formation. The Sleaford Member was deposited on a deeply channelled surface which can be correlated with the channelled base of Taylor's Upper Lincolnshire Limestone. However, the base of the overlying Clipsham Member is also strongly erosive in places. Only the basal Sproxton Member of Ashton's sequence can be readily recognised in the Peterborough District.

Table 1 Divisions of the Lincolnshire Limestone after Taylor (1963) and Ashton (1980)

Ashton, 1980 (Lincolnshire)		Taylor, 1963 (Northamptonshire)	
Upper	Clipsham Member *erosion*	coarse shelly and oolitic limestones with limestone pebbles	Upper Lincolnshire Limestone
	Sleaford Member *erosion*		
Middle	Lincoln Member *erosion*	?	'Crossi Beds' and Lower Lincolnshire Limestone
Lower	Greetwell Member	Pelletal and oolitic limestone	
	Sproxton Member	Sandy fissile limestones	

Taylor considered that the sandy base of the Lincolnshire Limestone was formed in part by the reworking of the underlying sandy Grantham Formation. The higher limestones were formed by the precipitation of carbonate in a shallow, current-swept sea. Ashton considered the sequence to represent the progression of an off-shore barrier complex across lagoonal and tidal flat deposits. His Lower Lincolnshire Limestone was of lagoonal origin and was succeeded by the 'agitated' deposits of his middle unit. These were in turn buried by the barrier-bar sediments which represent an environment of still higher energy.

GREAT OOLITE GROUP

The Great Oolite Group comprises the Upper Estuarine 'Series', the Blisworth Limestone, the Blisworth Clay and the Cornbrash. It ranges in age from the lower Bathonian to the lower Callovian. Figure 9 shows the variation in thickness in each part of the Group within the Peterborough District.

Upper Estuarine 'Series'

The Upper Estuarine 'Series' would more properly be called a formation. It crops out in the Nene valley from Castor to Peterborough and on both sides of the Marholm Fault from Marholm westwards. There is also a small outcrop on the north-east side of Helpston [1240 0563].

The formation is present at depth throughout the district, the thickness ranging from over 10 m in the north-west to less than 5 m in the south-east (Figure 9). A maximum thickness of 12.35 m was recorded in the Crowland No. 1 Borehole [2298 1067]. In the southern part of the district where the Lincolnshire Limestone is absent it is difficult to determine the lower boundary of the formation, and this was found particularly to be the case for the large number of site investigation boreholes drilled in the development of Peterborough (Horton and others, 1974, fig. 7).

The Upper Estuarine 'Series' is mainly argillaceous and was divided by Aslin (1968, p.234) into two parts which can be recognised in the Peterborough district. The lower part is the Lower Freshwater Sequence which appears to be devoid of marine fossils, but does contain rare chitinous debris. The sediments are mainly dark or brownish grey mudstones of seatearth type with abundant rootlets and listric surfaces. These may grade into silty mudstones or laminated sandy siltstones which are generally paler in colour. There are some sand-filled cracks, load casts and bioturbation at some horizons. The greatest recorded thicknesses were 5.34 m at Etton Waterworks Borehole [1412 0497] and 5.29 m at Parson Drove Borehole [3793 1052] but the division is thinner to the west and north-west of the district.

The Lower Freshwater Sequence is lithologically similar to the muddy facies of the Grantham Formation and, in the absence of the intervening Lincolnshire Limestone, it is possible that some beds attributed to the Grantham Formation may in fact be younger (Taylor, 1963; Aslin, 1968). In

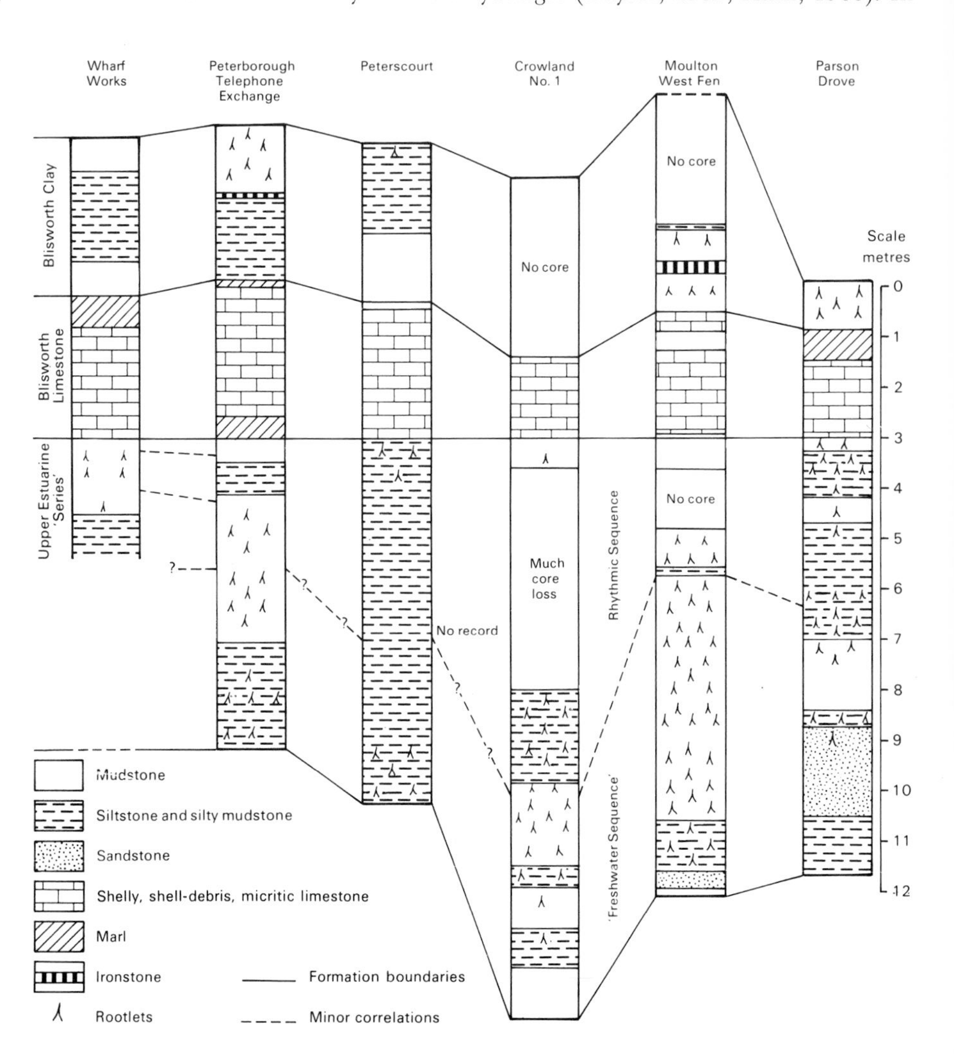

Figure 8 Lithological character of part of the Great Oolite Group

some areas an ironstone band is developed at the base of the sequence where it rests upon the Lincolnshire Limestone. At Elm Farm Borehole [1521 1291] a nodular ironstone occurs 0.1 m above the base and is overlain by a thin mudstone with limonite (iron oxide) grains. In the Crowland No. 1 Borehole sphaerosiderite occurs in the basal metre with a bed of ferruginous concretions marking the base. The ferruginous horizon has not been recorded south of the limit of the Lincolnshire Limestone.

Aslin termed the upper division of the 'Series' the Rhythmic Sequence which, to the west of the present district, comprises a cyclic sequence of marine sediments with brackish and fresh water intercalations. The marine beds are represented by shelly limestones and smooth mudstones, but the brackish water mudstones and silty mudstones contain a progressively restricted fauna which finally disappears in the fresh water part of the sequence, represented by beds with rootlets and seatearth textures. The typical pattern is not well defined in the Peterborough district where rootlets persist throughout the sequence and marine sediments are less clearly represented. The sediments include mudstones and silty mudstones, the marine beds being represented by shelly mudstone horizons.

The Rhythmic Sequence varies greatly in thickness from as little as 1.22 m in the Wharf Works Borehole [TL 1853 9797], Peterborough, to 7.91 m in the Crowland No. 1 Borehole.

Although limestones form a significant part of the Upper Estuarine 'Series' to the west of the district, they were recorded locally in only one borehole [1759 0136] north-west of Peterborough. Here the greater part of the formation comprises shelly limestones interbedded with shelly mudstones and calcareous siltstones.

The 'Series' is of lower to middle Bathonian age. The sedimentary character and restricted fauna of the Lower Freshwater Sequence indicates deposition in a marshy environment, probably crossed by rivers. The Rhythmic Sequence indicates subsequent minor invasions of the area by the sea.

Blisworth Limestone

The Blisworth Limestone crops out above the Upper Estuarine 'Series' along the slopes of the Nene valley from Castor to Peterborough, on the south side of the Marholm Fault around Marholm and to the west, and around Helpston. It is probably present at depth throughout the district and Figure 9 shows a maximum thickness of 4 m on the western margin and a thickness of less than 2 m eastwards from the Crowland No. 1 Borehole [2298 1067] to Wisbech No. 1 Borehole [4066 0842].

The formation ranges in age from middle to upper Bathonian and consists mainly of shelly limestone with beds of shelly marl and laminated shelly mudstone. Bivalve debris is the most abundant component of the rocks and is set in a microcrystalline matrix of calcite which probably represents a recrystallised carbonate mud. There is generally intense bioturbation. The base of the formation is usually sharp and is marked by a non-sequence above a bored surface of the underlying beds. The upper part of the Blisworth Limestone is shelly marl, forming a transition to the overlying Blisworth Clay.

The fauna is restricted and is dominated by *Liostrea hebridica*. The brachipod *Kallirhynchia sharpi* occurs at the base of the formation. The Blisworth Limestone marks a marine transgression which spread across the marshy flats of the

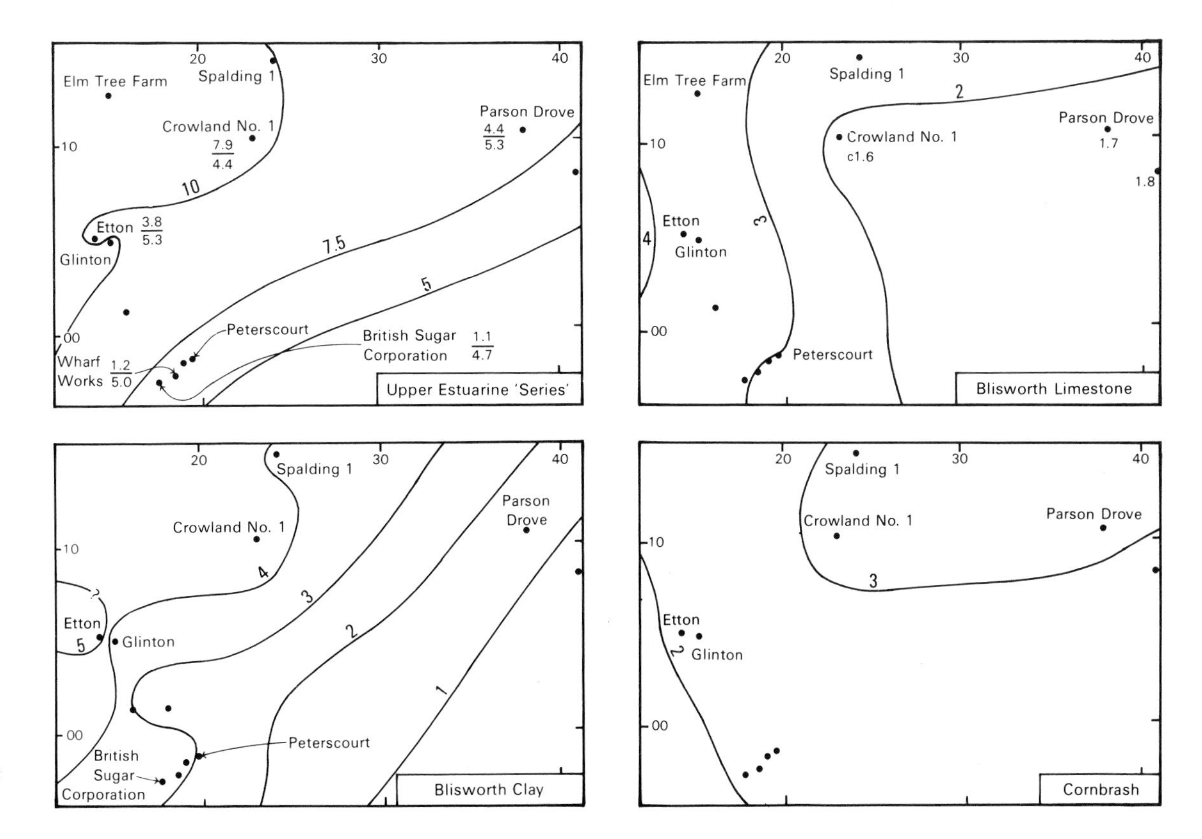

Figure 9 Variations in thickness within the Great Oolite Group

preceding Upper Estuarine period. The unsorted character of the sediments suggests that they were deposited in relatively shallow and generally quiet water.

Blisworth Clay

The main outcrop of the Blisworth Clay lies west of Peterborough between the River Nene and the Marholm Fault. There are more restricted outcrops to the south of the river and in the north around Helpston. The complex outcrop pattern in the north results partly from minor structures related to the Marholm Fault but it is also possibly due to superficial disturbances which have affected the overlying Cornbrash.

This clayey formation gives rise to a heavy brown soil which separates the clayey soil with brash of the Blisworth Limestone from the reddish brown brash soil of the weathered overlying Cornbrash.

Figure 9 shows the variation in thickness from 5.87 m at Etton Waterworks [1424 0518] to less than 1 m in the south-east part of the district, where it may possibly be absent locally. The Clay is about 3 m thick around Peterborough although greater and lesser figures were recorded in the site investigation boreholes for the city development, due possibly to difficulties in identifying the formation (Horton and others, 1974, fig.8).

The Blisworth Clay consists partly of dark grey and greyish brown mudstones, silty in their lower parts and with occasional rootlets. There are also more fissile mudstones with silty layers which are locally shelly. These latter may grade into marls and shelly limestones. Ironstones occur at several levels but have been recorded at only two horizons in recently examined boreholes or sections. Two bands were noted in the top 2.3 m of the formation in sections cut during the construction of the road interchange at Longthorpe [TL

Plate 1 Road cutting in the Great Oolite Group, near Thorpe Wood, Longthorpe.

The pale-coloured pockets just below the original surface level are cryoturbated masses of weathered flaggy Cornbrash limestone. The remainder of the cutting lies within the easily excavatable beds of the Blisworth Clay. The final excavation level is parallel to the top of the Blisworth Limestone, but 1 m below it. The machinery is ripping and pounding the massive limestones at the top of this formation, which breaks into blocks 5 × 4.5 × 0.4 m. (A 11653).

1562 9859] (Plate 1). Formerly, however, Porter (1861, p.92) described three ironstone nodule beds in a pit near Orton Waterville. The nodules which were up to 0.9 m in diameter occurred in the top 2 m of the formation. He also recorded an ironstone band, 0.2 m thick, immediately below the Cornbrash in a brick pit at New England (1861, p.91).

Fossils occur throughout the Blisworth Clay and include *Liostrea hebridica* and, in the higher beds, species of '*Corbula*'. The lack of coarse sediment and the presence of ironstone nodules probably led to Taylor's suggestion (1963, p.85) that in the Kettering district the Blisworth Clay was deposited in an 'enclosed basin or lagoon into which the sea rarely penetrated'. However, locally the lower part of the formation consists of mudstones and limestones with fossils indicative of a more open marine environment.

Cornbrash

The hard limestones of the Cornbrash form the higher ground west of Peterborough between the River Nene and the Marholm Fault. They give rise to a sloping bench in the western part of the city area because of a gentle south-easterly dip. To the east of the Peterborough Fault the Cornbrash is generally close to the surface beneath the Kellaways Beds or river gravels. South of the Nene the formation makes a slight feature defining the northern limit of the outcrop of the Kellaways and Oxford clays. There are also small outcrops north of the Marholm Fault to the south and south-east of Helpston.

Figure 9 shows the variation in thickness which ranges from over 3 m in the north-east to less than 2 m in the south-west of the district. In the Spalding No. 1 Borehole [2434 1478] geophysical records indicate a thickness of 1.8 m. In the south the evidence of site investigation boreholes in and around Peterborough indicates that the formation may be between 1.4 and 2.4 m thick (Horton and others, 1974, fig.10).

The Cornbrash consists mainly of limestone and weathers at outcrop to a reddish brown soil with pale brown limestone rubble. The rocks become increasingly hard and massive with depth particularly beneath a cover of the Kellaways Beds and Oxford Clay. They were formerly used for roadstone and were exposed in many shallow pits throughout the outcrop but during the last survey only two significant exposures were found. An overgrown section on the side of the Oundle Road [TL 1528 9666], north-west of Orton Waterville, showed 0.9 m of shelly limestone with numerous clay-filled burrows. Fossils collected from the surface spoil indicate that this belongs to the upper part of the formation. In the shallow cutting [TL 1836 9995] north of Peterborough railway station there were exposures of up to 2 m of weathered rubbly limestone.

Most of the more recent data on the formation within the district comes from site investigation boreholes (Horton and others, 1974) or from holes drilled for schemes for filling the old brickpits (Callomon, 1968). The limestones are bluish grey in the unweathered state and are typically made up of fine shell debris, generally less than 1 mm across, in a microcrystalline calcite matrix. They are intensely bioturbated and contain clay-filled burrows. At some levels they may pass into marls or calcareous mudstones with increasing clay content. The most persistent argillaceous horizon marks the top of the Cornbrash and passes upwards into the overlying Kellaways Clay. The base of the formation is abrupt and rests with a non-sequence on a commonly bored surface of the Blisworth Clay.

The fauna is dominated by brachiopods and bivalves with which are associated ammonites, echinoids and serpulids. The Cornbrash has been separated into Lower and Upper divisions both of which have been recognised within the district. The Lower Cornbrash lies in the *Clydoniceras discus* Zone and is therefore of upper Bathonian age (Cope and others, 1980b, fig.6b). The Upper Cornbrash, proved in boreholes and clay pits immediately south of the district, probably lies within the *Macrocephalites macrocephalus* Zone of the basal part of the Callovian Stage (Cope and others, 1980 b, fig.8). Callomon (1956, p.251) showed that the top of the Cornbrash is diachronous and that the deposition of the overlying Kellaways Clay started earlier in the south of the district than to the north.

The Cornbrash is entirely marine. It accumulated in a shallow current-swept sea, where the sediment was repeatedly reworked both by organisms and by currents. Partial cementation occurred in places and subsequent erosion produced pebbles which were recycled into younger deposits. One such period of erosion is thought to have marked the transgression which preceded the deposition of the Upper Cornbrash.

ANCHOLME GROUP

The term Ancholme Group was established from mapping in the Brigg area of south Humberside to cover the higher Jurassic formations which are mainly clays. Southwards across Lincolnshire these formations cannot satisfactorily be mapped separately, partly because of a lack of distinctive lithological horizons, but mainly because the low-lying outcrops of the clay formations are covered by thick and extensive drift deposits.

The Group includes the Kellaways Beds, Oxford Clay, West Walton Beds, Ampthill Clay and Kimmeridge Clay. Only the lower three of these occur in the Peterborough district.

Kellaways Beds

The Kellaways Beds comprise the Kellaways Clay and Kellaways Sand and are of lower Callovian age. They crop out on the south side of the Nene valley from Orton Waterville eastwards to New Fletton and thence through the eastern, central and northern parts of Peterborough to the Marholm Fault. They are largely obscured by drift deposits in the area of the city. North of the Fault the Kellaways Beds crop out to the south and south-east of Helpston and are present beneath Fen drift to the north as far as Market Deeping and to the north of Langtoft in the north-west corner of the district.

Kellaways Clay

The Kellaways Clay is a grey mudstone with scattered phosphatic nodules. It is relatively poorly fossiliferous with scattered bivalves and ammonites. Sections in the Peterborough brick pits have yielded ammonites which indicate that the Clay lies within the *kamptus* Subzone of the basal Callovian *macrocephalus* Zone (Callomon, 1968). This author also suggested a thickness of about 2.1 m for the Kellaways Clay in the pits south of Peterborough while the Parson Drove [3793 1052] and Tydd St Mary [4307 1737] boreholes proved 1.7 and 2.1 m respectively.

Kellaways Sand

The Kellaways Sand consists of grey fine-grained sandstones and silty sandstones, interbedded with siltstones and silty mudstones. The beds are generally laminated but are also intensely bioturbated, with burrows filled with silty sand. Thick-shelled bivalves and shell debris occur throughout. There is a transition upwards from the Kellaways Clay with increasing sand content. The upper boundary is more clearly defined and is marked, in the Parson Drove Borehole, by a shelly pyritic layer at the base of the overlying Oxford Clay. The Kellaways Sand was proved to be 2.86 m thick in the Parson Drove Borehole and 2.5 m in the Tydd St Mary Borehole. Callomon (1968, p.284) recorded ammonites of the *koenigi* and *calloviense* subzones, of the lower Callovian *calloviense* Zone, from the pits south of Peterborough.

Oxford Clay

The Oxford Clay is the thickest and most widespread of the Jurassic formations which crop out within the district. The most extensive outcrops are in the Old Fletton area south of the Nene, around Eye [230 030] and Thorney [285 035], and

Plate 2 Kings Dyke Brick Pit, Whittlesey [TL 245 975].

Shale-planar excavator digging the Lower Oxford Clay for brick-making. The section illustrates the marked stratification with alternations of darker greyish brown bituminous mudstone and paler mudstone. The excavator works down to the lowest nodule bed of the Oxford Clay. The overburden comprising up to 5 m of Flandrian and Pleistocene deposits, and weathered Oxford Clay (the callow) is removed prior to extraction of the brick clay. (MN 26846).

to the west of Werrington [167 030], north of the Marholm Fault. It underlies the Fen deposits over much of the district and crops out along the margins of the Fenland 'islands', where the March Gravels produce features which rise above the level of the Fens. In these areas it crops out on the slopes below the gravel remnants, for example around Coates [TL 306 977], Whittlesey [TL 270 970], Eye, The Reaches ridge [240 040] and south-west of Thorney. At outcrop, for example near Werrington, it forms fairly flat, poorly drained ground.

The Oxford Clay has been separated into Lower, Middle and Upper divisions, the first two of which are of Callovian age while the topmost is of lower Oxfordian age.

Lower Oxford Clay

The Lower Oxford Clay crops out above the Kellaways Sand in the Helpston–Gunthorpe and Old Fletton areas. It consists of brown fissile mudstone interbedded with pale grey blocky mudstone. The former is shelly and bituminous whilst the latter is much less fossiliferous. Thin shell beds occur throughout, and these may be composed mainly of one species, such as *Nucula*, or of crushed ammonites. They mark pauses in deposition and, in conjunction with burrowed horizons, may be used to define small-scale rhythms of sedimentation. There is a band of large calcareous septarian nodules in the lower part of the division, and near the top another calcareous band is known as the Acutistriatum Band-Comptoni Bed. The Acutistriatum Band is a calcareous mudstone which passes laterally into a lenticular limestone. It was 0.59 m thick in the Parson Drove Borehole [3793 1052] and overlies the Comptoni Bed which comprises shelly marls and mudstones with burrows. The brick pits in the Lower Oxford Clay (Plate 2) have provided a rich fauna which includes ammonites, bivalves, gastropods, belemnites, crustaceans, fish and reptiles. The stratigraphical distribution of the ammonites was described by Callomon (1968, p.271). The Lower division spans the upper *calloviense, jason, coronatum* and lower *athleta* zones of the Callovian.

Callomon (1968) suggested a thickness of 16.61 m for the Lower Oxford Clay in the Peterborough brick pit sections and this compares with 18.07 m measured in the Parson Drove Borehole.

Middle Oxford Clay

The Middle Oxford Clay probably lies beneath the Fen deposits in a belt trending south-eastwards from Hop Pole [190 140] to the west of Crowland and thence towards The Reaches ridge and Eye Green [230 040]. It was formerly exposed in the disused Eye Green brick pit [230 034]. South of the Marholm Fault it extends from a point just east of Eye to Chicell's Hurst [270 030] and Hill Farm [279 026] and thence southwards to Whittlesey. It is also exposed in the brick pits west of Whittlesey. Drilling for sand and gravel in the Eye area yielded specimens of Oxford Clay from the bottoms of the boreholes (Booth, 1982, 1983) which gave microfaunal evidence of the presence of the middle division.

The Middle Oxford Clay consists of grey, slightly fissile mudstones with abundant immature bivalves and with fine shell debris at some levels. The lowest beds are much burrowed. A calcareous bed, which marks the top of the division, is correlated with the Lamberti Limestone of the ground to the south of the district. The Middle Oxford Clay is less fossiliferous than the lower division and many of the ammonites are immature, although species of *Kosmoceras* persist and indicate an upper Callovian age. A calcareous worm, *Genicularia vertebralis*, is a fossil of the topmost Lower and basal Middle Oxford Clay and, being resistant to erosion, forms a useful stratigraphical marker in mapping from field debris. The Parson Drove Borehole [3793 1052] proved 20.6 m of the middle division.

Upper Oxford Clay

The base of the Upper Oxford Clay is thought to run beneath the Fen drift in an approximately north–south line through Crowland [240 100] towards Middle West Farm [257 044] near the Marholm Fault. There is an outcrop west of Cat's Water Farm [247 040] which is fault-bounded to the north and south. South of the Marholm Fault there are outcrops south-west of Thorney [282 042] and the lower boundary probably extends to the vicinity of Coates [TL 306 977]. The upper boundary with the West Walton Beds is shown conjecturally on the geological map from north of Gedney Hill [340 115] to the west of Guyhirn [390 030] and south towards West Fen [TL 373 960]. The Oxford Clay was cored in the Parson Drove Borehole where the upper division was reckoned to be about 26 m thick.

The Upper Oxford Clay is a pale grey, blocky mudstone with rhythmic units, defined by burrowed carbonaceous horizons which are thought to represent pauses in sedimentation. It is less fossiliferous than the two lower divisions and the fauna is characterised by bivalves including large specimens of *Gryphaea dilatata*. The ammonite fauna is dominated by cardioceratids which indicate a lower Oxfordian age.

West Walton Beds

The geological map shows the Oxford Clay overlain by the Corallian, the latter term being formerly widely used elsewhere to include calcareous strata which in this area are largely represented by the argillaceous West Walton Beds and the Ampthill Clay. Only the West Walton Beds have been proved within the Peterborough district, although from borehole evidence to the north-east of the district it is possible that the Ampthill Clay may underlie the Fen drift in the north-east corner.

There is a restricted outcrop of the West Walton Beds, fault-bounded to the north and south, to the east of Cat's Water Farm [247 040]. The basal strata (labelled Cr on the geological map), exposed in a ditch [2455 0406], comprise about 0.8 m of weathered, shelly, sandy limestone. A borehole [2458 0408] sited close to the junction with the Oxford Clay proved rust-mottled, grey clay with sandy partings and calcareous concretions. To the east another hole [2500 0384] penetrated terrace gravels to prove clays with a foraminiferal fauna indicative of the middle Oxfordian *densiplicatum* Zone. These are thought to overlie the basal calcareous beds.

The main area of occurrence of West Walton Beds lies

beneath the Fen drift along the eastern margin of the district, the junction with the Oxford Clay being drawn conjecturally. In this area the strata are known only from boreholes. The Parson Drove Borehole [3793 1052] proved 10.6 m of grey silty mudstone, ascribed to the West Walton Beds, resting upon mudstone of the Upper Oxford Clay. The sediments contain rhythmic sequences, consisting ideally of dark, carbonaceous, silty mudstone, with abundant plant and shell debris, which grades upwards into pale grey, calcareous, silty mudstones, siltstones or silty limestones. This sequence may be interrupted by burrowed horizons which represent periods of slight erosion or non-deposition and which mark the beginning of a new rhythm. These lithological sequences and faunal changes were used to establish a standard Fenland sequence for the West Walton Beds (Gallois, 1979; Gallois and Cox, 1977). Beds 1 to 15 of that sequence were recognised in the Parson Drove Borehole, although strata below Bed 7 are transitional to the underlying Upper Oxford Clay (Horton and Horrell, 1971) which make it difficult to define the base of the West Walton Beds on lithological grounds. A small subcrop of West Walton Beds is present beneath the Fen drift near Gull House [2458 1470]. Its presence is inferred from the recognition of the geophysical signature of the West Walton Beds in the logs of the Spalding No. 1 Borehole.

FIVE

Pleistocene deposits

The Quaternary sequence of the Peterborough district comprises glacial and interglacial Pleistocene deposits which are overlain by Flandrian sediments. The latter are the most widespread as a relatively thin sheet which forms most of the low-lying Fenlands. The Pleistocene includes glacial lake deposits, till, glacial sand and gravel, fluvial gravel terraces and contemporaneous marine gravels.

GLACIAL LAKE DEPOSITS

Glacial lake deposits occur south of Stanground [TL 210 965] on the southern margin of the district. They are thought to fill a shallow depression which is up to 1.5 km wide and extends south of the district to the vicinity of Farcet where the deposits are at least 15 m thick and were formerly exposed (Horton and others, 1974, p.51). The sediments comprise brown and grey clays, thinly interbedded with paler clays, silts and fine sand. There are occasional beds of pebbly clay. Lacustrine deposits have been proved beneath the Fen deposits in boreholes in the north-east of the district around Sutton St Edmund, Parson Drove and near Guyhirn. They are up to about 2 m thick and lie above till.

TILL

Till crops out on the western margin of the district, north of Castor, on the southern edge, south-east of Old Fletton, and in the south-east corner around Westry. Boreholes indicate that it is the most wide-spread member of an extensive sheet of glacial drift which underlies the Fen deposits north and east of a line from Crowland [240 100] through Guyhirn [400 040] to Westry (Figure 10).

The till is chalky boulder clay which is a grey stiff clay with varying proportions of silt and sand, and containing erratic stones. These latter include chalk and flint, abraded Jurassic fossils, rounded pebbles of quartz and quartzite, thought to be derived from Triassic conglomerates, and a variety of other far-travelled rocks (Sabine, 1949). The till weathers to pale brownish grey and may be decalcified in the topmost layers. It is considered to be a product of the Anglian glaciation and is correlated with similar tills of the King's Lynn area and the Midlands.

The till is at least 4.4 m thick at a bridge [3832 1096] south-west of Harold's Bridge and was proved at over 12 m in the ground north-east of the district. There may be at least 5 m present in the Westry area. The outcrops north of Castor are remnants of a formerly more extensive deposit.

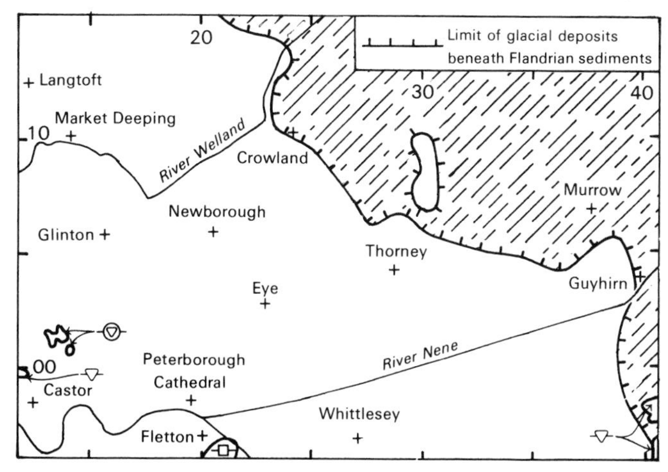

a. Glacial deposits

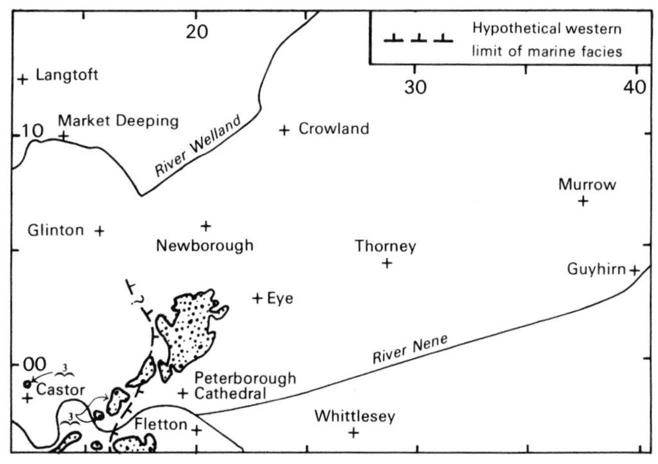

b. Third Terrace deposits and Woodston Beds

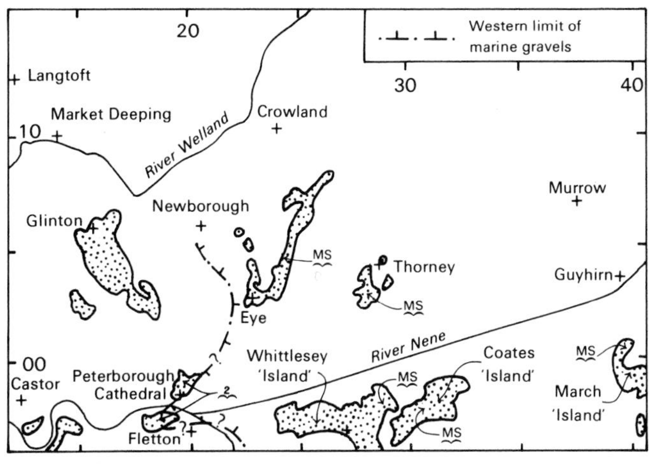

c. Second Terrace deposits and March Gravels (MS)

Figure 10 Extent of glacial deposits, older river gravels and marine gravels

GLACIAL SAND AND GRAVEL

There are deposits of glacial sand and gravel around Belsize Farm [138 011], south-west of Marholm, and there is a small area [TL 216 963] south-east of Stanground. These sediments consist of well to poorly sorted gravels and sand with thin bands and lenses of silt and clay. The gravel fraction includes flint, chalk, quartzite, quartz, Jurassic limestone and ironstone, together with ironstone probably from the Cretaceous Carstone, and derived fossils, particularly the robust shells of *Gryphaea*. Boreholes north and north-east of Thorney proved glacial gravels (Booth, 1982) with thicknesses ranging from 0.3 to 2.7 m. It is difficult to recognise glacial gravel in borehole samples but the presence of chalk and a relatively low content of local rocks probably distinguishes the glacial deposits from the younger gravels.

GLACIAL HISTORY

It is thought that during the Anglian glacial period (at least 200 thousand years ago) an ice sheet advanced from the north across a low-lying Fenland basin, the south-west boundary of which in the Peterborough area lay close to its present line. This basin probably included the valleys of a pre-glacial drainage system which, in the Terrington–Denver area, descend to at least 70 m below OD (Gallois, 1974, 1979), and in the mouth of The Wash lie at about 100 m below OD. Boreholes show that south-east of Crowland [240 100] the sub-glacial surface is between 4 and 7 m below OD. It is deeper in the eastern part of the district, being more than 20 m below OD at Poplar Tree Farm [4056 1343], 27 m below OD in the Parson Drove Borehole [3793 1052] and up to about 13 m below OD at Guyhirn [400 040].

The ice sheet laid down glacial deposits, probably mainly boulder clay, on the basin floor, filling the local depressions, and these deposits were also left on the higher ground as the ice advanced to the south-west. At this time finer deposits accumulated in pro-glacial lakes. In the later stage of the glaciation melt water would have deposited glacial gravels, probably along the ice front or in valleys, and finer lacustrine sediments were laid down in hollows which had been dammed up by standing ice or by ice-derived debris.

It is probable that, originally, the whole district was virtually covered with glacial deposits but erosion, in the period prior to the Flandrian, removed most of them in the western part to leave the relict distribution indicated in Figure 10a.

RIVER AND MARINE GRAVEL DEPOSITS

Gravel deposits occur throughout the western and southern parts of the district, the First Terrace gravels of the Welland and Nene being the most widespread. Most of the deposits are of non-glacial origin and were formerly grouped together as 'Fen gravel' by Skertchly (1877) who described the gravels of the Fens proper as 'undoubtedly marine', and who also recognised that the gravels along the main rivers and the spreads on the Fen margins passed laterally into the marine deposits of the central Fens. Castleden (1980) was of the same opinion in linking the second and third terraces of the Nene with the gravel-capped Fenland 'islands' at Whittlesey and Westry.

The geological survey of the district and borehole data from sand and gravel resource surveys have confirmed these views and Booth (1981, 1982, 1983) showed that the river and marine gravels are broadly similar in composition. The correlation of the fluviatile and marine deposits is illustrated in Table 2.

Table 2 Correlation of Terrace deposits and Marine equivalents

River terrace deposits		Marine deposits
Welland	Nene	
1st	1st	Abbey Gravels
2nd	2nd	March Gravels
—	3rd	Woodston Beds

The gravels accumulated after the Anglian glaciation as a result of erosion and redeposition of the glacial drift and of the Jurassic country rocks as they progressively became exposed by denudation. The dominant components of the gravels are Jurassic limestone and flint derived from the glacial deposits. Other rock types include quartzite, quartz, ironstone and sandstone, together with derived Jurassic fossils such as *Gryphaea*.

Third terrace deposits (Nene) and Woodston Beds

Gravelly deposits of the Third Terrace of the Nene lie in a belt extending from the west of Orton Waterville [TL 145 965] north-eastwards through Longthorpe [TL 163 985] across Peterborough to the vicinity of Dogsthorpe [200 020] (Figure 10b). Borehole data indicate that the height of the base of these deposits varies from about 18 m to about 1 m above OD with a general fall to the north-east. The gravels are generally less than 2 m thick in the vicinity of Orton Waterville but are between 3 and 4 m thick immediately south of the district at Orton Longueville and Woodston. The thicker sequences may include silty clays which are probably equivalent to the Woodston Beds. Trench exposures in the southern area show that the thin gravel deposits are intensely cryoturbated (Horton and others, 1974, p.53). In Westwood [TL 1755 9931], a temporary section showed about 0.7 m of loam and sandy clay upon at least 1.2 m of poorly sorted gravel.

The Woodston Beds are mainly silty clays and fine sands and in the type area, immediately south of the district at Hicks Brick Yard [TL 190 956], they interdigitate with gravels. Farther west, as at Orton Longueville, they occur below gravels which have been mapped as Third Terrace. The fauna of the Woodston Beds includes marine and freshwater shells and fossil bones. Clay horizons have yielded pollen which indicate a correlation with the Hoxnian interglacial. The fauna and flora suggest deposition in a warm temperate climate in an environment ranging from estuarine through salt marsh to fluviatile or possibly ponded. In contrast to this, Castleden (1980, p.34) cited involuted structures as evidence of contemporaneous periglacial conditions, but it is also possible that these structures result from post-depositional cryoturbation (Horton and others, 1974).

Second Terrace deposits (Nene and Welland)

The Second Terrace of the Nene occurs on the valley sides downstream from the western margin of the district and forms benches around Peterborough Cathedral [TL 194 987], at Stanground [TL 209 970] and around Oxney House [225 009] (Figure 10c). The terrace form is best developed near the Fen margin, and the top and base of the deposit show a general fall towards the Fens, the upper surface ranging in height from about 16 m to about 5 m above OD, and the base from about 13 m to about 3 m above OD. The terrace deposits are generally not more than 2 m thick, although in the Woodston area a temporary section [TL 1842 9735] showed 7.6 m of clean limestone gravel and boreholes proved up to 4.6 m of gravel. These greater thicknesses may reflect the presence of a gravel-filled channel beneath the terrace.

The deposits consist largely of limestone gravel and minor quantities of flint and other pebbles (Booth, 1981, 1982). Beds of sand, silt or clay are not common. Boreholes in the Second Terrace [TL 1496 9677] north-west of Orton Waterville proved a downward sequence of 3.5 m gravel; 0.7 m sand and silt; 1.55 m humic clay; 0.25 m gravel. The humic clay contained freshwater ostracods, one of which, *Cytherissa lacustrica* (Sars), was first recorded from cold, deep Norwegian lakes and may indicate a sub-Arctic environment. Kennard and Woodward (1922, p.128) collected freshwater and terrestrial molluscs in association with marine shells from the terrace at Woodston. This terrace may thus be equivalent in part to the March Gravels.

The Second Terrace of the Welland is represented by a deposit north of Marholm [150 030] and a more extensive spread which extends from Gunthorpe [183 025] north-westwards to Glinton [154 060] and Peakirk [170 066]. Here there is also a general slope towards the Fens with the top of the Terrace ranging from about 12 to 7.5 m above OD and the base ranging from about 10.7 m to about 4 m above OD.

March Gravels

Within the Peterborough district the March Gravels cap 'islands' in the Fens, the main areas lying north-east of Eye, south-west of Thorney, around Whittlesey and Coates, and at Westry in the south-east corner of the district (Figure 10c). The last-named locality is a part of the March 'island' from which Baden-Powell (1934, p.193) introduced the name. At Whittlesey they form a bench which descends from about 7.6 m above OD in the west to about 5 m above OD in the east. Boreholes proved that the deposits vary from sandy gravel to clayey, pebbly sand and that the thickness ranges from 2.1 to 5 m. The base is at about 2.8 m above OD in the west and between 1.1 and 3.2 m above OD in the east. Up to 4 m of gravel with pebbles of limestone, flint and ironstone are exposed as overburden in the clay pits west of Whittlesey. The base of the Gravels can be projected westwards from Whittlesey to correlate with that of the Second Terrace of the Nene (Castleden, 1980, p.38; Booth, 1981, p.5). Frost wedges were described by Skertchly (1877, fig.20) and Castleden (1980, fig.3), the evidence illustrated by the latter suggesting the incoming of periglacial conditions during the period of deposition of the March Gravels.

The Coates 'island' forms a bench at about 6 m above OD with the underlying Oxford Clay cropping out locally around the margin, as around Whittlesey. Limited borehole data indicates that the base of the Gravels lies at about 2 m above OD and that their thickness ranges up to at least 3 m. At Westry the surface of the Gravels lies at between 3.7 and 4.2 m above OD.

The March Gravels cap two ridges which extend northwards from Eye [230 030]. The surface lies at about 6.5 m above OD in the south and descends to about 3 m above OD in the north. The base of the deposit descends from about 4 m above OD in Eye to 3.1 m below OD near Northolm Farm [2295 0427] and to 2 m below OD near St Vincent's Cross [2579 0745]. The Gravels comprise clayey, sandy gravel and clayey, pebbly sand with flint and limestone as the main components of the gravel fraction. The thickness ranges from 2.3 to 8 m. Booth (1982, p.52) noted that in this area the basal beds are mainly fine sand with abundant marine and estuarine shells, probably filling local scour channels.

Marine fossils were recorded from the Eye area by Marr and King (1928) in association with mammalian bones and driftwood. The fauna described by Baden-Powell (1934), and that collected during the geological survey, includes bivalves and gastropods which indicate an intertidal, shallow water environment which may have become brackish locally. Although bivalves such as *Macoma calcarea* indicate cold water, the bulk of the fauna suggests temperate, shallow, freshwater and estuarine conditions. It is considered that the March Gravels may be of Ipswichian age, but the presence of frost wedges within the gravel in places, marks the onset of periglacial conditions during their deposition.

First Terrace deposits (Nene and Welland)

Gravel and sand deposits of the First Terrace lie on either side of the Nene downstream from the western margin of the district and spread out east of Peterborough as fan-shaped areas on the edge of the Fens. The First Terrace of the Welland lies mainly to the north of that river and covers the greater part of the north-west corner of the district. Figure 11 shows the areas of outcrop, contours on the base of the deposits, the courses of underlying buried channels and the western limit of the contemporaneous area of marine deposits. These latter are represented by the Abbey Gravels. The Terrace surface marks the upper limit of aggradation of the deposits which extend beneath the river alluvium and the Flandrian deposits of the Fens.

In the Nene valley the Terrace surface descends from about 8 m above OD on the western margin of the district to about 1 m above OD near Thorney [277 053]. West of Peterborough the gravels fill a buried channel which is deepest and narrowest near Peterborough Bridge [TL 1916 9818] where the more resistant Cornbrash has produced a marked narrowing of the valley. West of the city thicknesses are

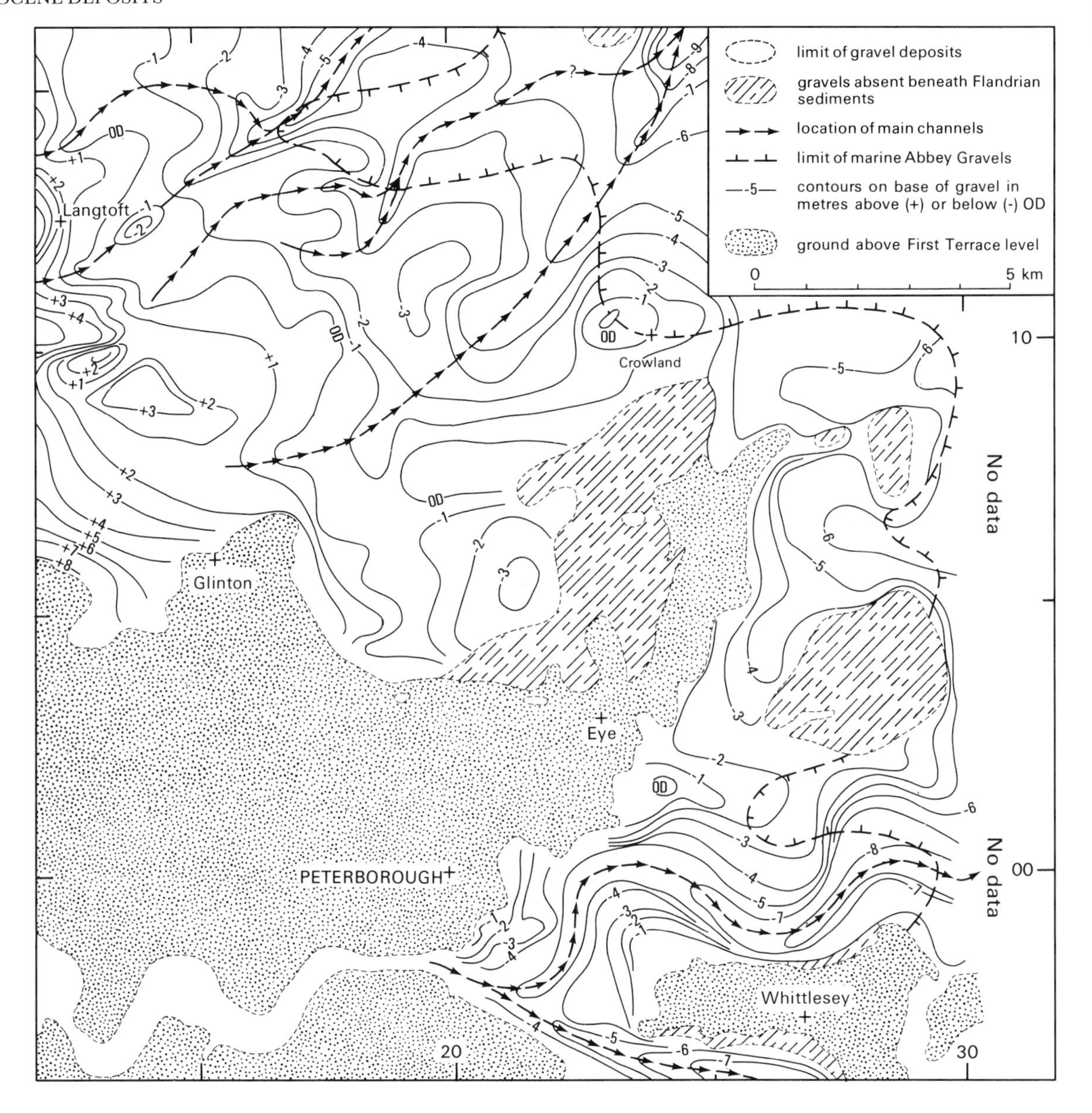

Figure 11 Extent of First Terrace deposits and Abbey Gravels

generally less than 5 m but figures of up to 11 m have been recorded, for example in the south-west corner of the district at [TL 1197 9665].

Where the river entered the Fen lowlands, east of Peterborough, broad spreads of gravel were deposited and boreholes show that the lithology changes from well sorted gravel to clayey, pebbly sand in a seaward direction. The sub-terrace channel also extends to the east and bifurcates north of Stanground with a branch on each side of the Whittlesey 'island'. The more northerly channel descends to at least 9.2 m below OD north-east of Bassenhally Farm [TL 2959 9962] and possibly 13.3 m below OD near Adventurers' Land [3458 0148].

On the eastern side of Peterborough, between Fengate and Newark, the upper surface of the Terrace falls from about 6 m above OD at the western edge to between 3 and 4 m above OD near the Fen edge. Baden-Powell (1934, p.194) listed deposits in this area as March Gravels but the more recent geological survey found no marine fossils in temporary sections in the area.

There is a wide spread of First Terrace gravels between Eye [230 030] and Thorney [280 040] and the upper surface descends from about 4 m above OD at the southern end to about 1 m above OD near the northern end. The Terrace is covered by Flandrian deposits to the north where boreholes showed that the base of the Terrace deposits has descended to as low as 6.6 m below OD at a locality [2845 0655] north of Lodge Farm.

The north-west part of the district is largely covered by a spread of First Terrace deposits of the Welland. These extend westward along the river valley almost to Stamford and were laid down as a broad fan where the river emerged into the Fen lowlands. Within the district the Terrace has a maximum height of 11 m above OD [1191 0798] west of Maxey, whence it slopes to the north and east to pass beneath Flandrian deposits at between 2 and 3 m above OD.

The channels beneath the First Terrace of the Welland (Figure 11) are not related to the modern canalised course of the river. They are complex and are linked to buried tributary valleys. The gravel fill descends to at least 6.2 m below OD [1616 1387]. Several boreholes recorded a thickness of over 5 m with a maximum figure of 7 m. The deposits become thinner away from where the Welland and the adjacent rivers enter the Fens, which led Booth (1982, p.8) to conclude that they were formed as alluvial fans. In a seaward direction the gravels pass into gravelly loamy sand.

In some places the youngest beds are sands, silts and clays with occasional freshwater molluscs. A thickness of 1.7 m was recorded near Northborough [1545 0768] and west of Etton waterworks [1390 0481]. These sediments may be overbank deposits laid down during the final period of aggradation of the Terrace.

From Peakirk [170 065] to Crowland [240 105] First Terrace deposits and Abbey Gravels from a sinuous outcrop. This 'Crowland Ridge', however, only rises locally above the surrounding Fen deposits and, despite its form, is not a former channel as suggested by Prentice (1950, p.136) and Booth (1982). In fact, south of the Decoy [200 075] the outcrop coincides partly with a sub-gravel ridge rising to at least 1 m above OD. The First Terrace/Abbey Gravels deposits may extend to the north-east beneath the Fen sediments although, to the south of Crowland, they appear to be largely absent, with the Fen deposits resting directly upon Oxford Clay.

Silt and clay horizons in the lower part of the First Terrace deposits have yielded fossils at Baston Pit [126 143] which include gastropods and bivalves indicative of a climate of arctic or sub-arctic intensity. These are associated with a fauna of beetles which include boreal species. The evidence of a cold environment is supported by the presence of frost wedges within the Terrace sediments. Some were formed prior to the deposition of the fossiliferous horizons while a second set were formed at some later stage (Booth, 1983, pl.1).

Silt lenses in the basal deposits at the Maxey gravel pit [134 076] yielded a predominantly freshwater shelly fauna, all the species of which are represented in modern British faunas. These thus suggest more temperate conditions than those indicated at Baston.

A section at Tanholt Farm gravel pit [2395 0137] in the First Terrace of the Nene, showed two clayey lenses in the lower part of the deposit and two sets of frost wedges, the older having developed in the period between the deposition of the clayey lenses. South-west of the district, along the Nene valley, mammalian bones from the base of the First Terrace in the Northampton area have given radiocarbon ages of 37 000 to 42 000 years BP (Brown, 1967), while a tundra-type assemblage of plants and beetles gave an age of 28 225 ± 330 years years BP (Morgan, 1969). This evidence led Castleden (1976) to conclude that the First Terrace deposits of the Nene accumulated during a late Devensian cold phase.

The above evidence of a cold climate during the deposition of the First Terrace deposits is at variance with the findings of French (1982) from the Maxey pits. From an excavation in the pits, not located in detail, French collected a fauna and flora from two fossiliferous horizons within 1.1 m of sediments which filled a channel enclosed by the Terrace gravels. West (*in* French, 1982) suggested that the pollen in the lower horizon might indicate an Ipswichian Ib age while that of the upper horizon might be of Ipswichian IIb age. The molluscan fauna belonged to a wide spectrum of freshwater and land habitats. French concluded that the presence of two temperate species of mollusc, which no longer live in Britain, and the existence of oak and hazel pollen indicated deposition during the warmest part of the Ipswichian interglacial period.

Abbey Gravels

The Abbey Gravels lie at the north-eastern end of the sinuous outcrop which extends from Peakirk [170 065] to Crowland [240 105]. They are clayey, sandy gravels, locally with a marine fauna, and resemble the deposits of the First Terrace with which they are contiguous. The only other outcrop of the Abbey Gravels is at a locality [320 050] east-north-east of Thorney. The Abbey Gravels have a similar lithological composition to that of the March Gravels but occur at a lower level.

Skertchly (1877, p.202) first noted the marine fauna and recorded *Macoma* (formerly *Tellina*) *balthica* and *Turritella communis*. Booth (1982) noted that marine shells generally occur in the more sandy beds which lie at the base of the Gravels. Figure 11 shows the outcrops of the Abbey Gravels and the approximate western limit of marine sedimentation in First Terrace times. This limit is of course based largely on borehole data and in fact represents a zone within which the fluviatile and marine facies interdigitate.

CROWLAND BED

The Crowland Bed is the name given to the widespread layer which separates the basal member of the Flandrian sequence, the Lower Peat or the Barroway Drove Beds, from the underlying original land surface. It is generally less than 1 m thick, structureless and commonly contains rootlets. It generally occurs at or below the watertable and is therefore unoxidised and consists of mottled grey silty clay or clay with scattered pebbles of fresh, dark flint and limestone. The matrix resembles the underlying deposit, but contains elements derived from adjacent areas, presumably by solifluction, although the extremely gentle slope of the ground limited this process. The pebbles are probably relict from pre-existing deposits and have been incorporated into the Crowland Bed by repeated cryoturbation. In time, it formed the bed upon which the Lower Peat developed and now shows characters associated with soil genesis.

Gallois (1979) recognised a comparable unit below the Lower Peat in the Wisbech–King's Lynn district. He concluded that it might be partly a head deposit, partly a freshwater channel sediment or partly a marine lag deposit.

HEAD

Head deposits are widespread, but are generally too thin or insufficiently extensive to show on the map. Those mapped can be grouped into those which occur in valley floors, such as the deposits near Castor Hanglands [119 016], south of Helpston [119 037], at Paston [186 022] and west of Longthorpe [TL 154 984], and those south of Castor [TL 129 974], north of Orton Waterville [TL 157 966] and south of Woodston [TL 180 962] which blanket valley slopes. The deposits were formed locally from indigenous material as a result of soil creep or solifluction, though flooding may have occurred during the accumulation of the valley floor deposits.

POST-ANGLIAN PLEISTOCENE HISTORY

The low-lying Fenland basin was probably in existence in Middle Pleistocene times. The history of the Anglian glacial deposits has been described above (p.18). The river terraces and their marine equivalents are the local representatives of deposition in the subsequent part of the Pleistocene period. They are considered to have accumulated by the reworking of the glacial sediments and of the Jurassic country rocks as these became increasingly exposed by spasmodic denudation.

The Third Terrace of the Nene and its marine equivalent, the Woodston Beds, are thought to be of Hoxnian age, from the presence of pollen in the latter which are considered to indicate a warm temperate environment. The pebble content and the presence of derived Chalk foraminifera in the Woodston Beds indicate that these are younger than the glacial deposits.

A period of erosion preceded the deposition of the Second Terrace whose base lies about 3 to 4 m below the thalweg of the Third Terrace. The Second Terrace deposits have yielded ostracods (p.19) which may indicate a subarctic environment, and frost wedges in the marine equivalent March Gravels at Whittlesey have been taken as evidence of contemporaneous periglacial conditions. In contrast, the fauna in the March Gravels at Eye may suggest a temperate climate although one or two of the bivalves indicate cold water conditions. The balance of evidence for this period seems to favour cold to temperate conditions and a correlation has been made by Mitchell and others (1973, table 2) with the early Ipswichian.

Another period of downcutting and planation preceded the deposition of the First Terrace gravels and the marine Abbey Gravels (Figure 12). The aggradation of the terrace spreads and the formation of coalescing alluvial fans indicates adjustment to a relatively high sea level. Radiocarbon ages of 37 000 to 42 000 years for bones from the Northampton area, and of about 28 200 years for beetles from the same area (*see* p.21) suggest an age for the Terrace covering most of the Middle Devensian. It has to be noted that this date does not tally with the Ipswichian age suggested by the fauna and flora collected from the Maxey area by French (*see* p.21). At all events the First Terrace and Abbey Gravels are considered to be older than the Hunstanton Till which was deposited in The Wash, and on the north Norfolk and Lincolnshire coasts, in the Late Devensian. This ice sheet must surely have blocked the Fenland Basin creating a pro-glacial lake. However, there is no evidence locally of sedimentation associated with this ice sheet, nor of the effects of the 100 m fall in sea level resulting from this glaciation.

Throughout the Post-Anglian period the boundary of the Fenland basin has remained almost the same. Over much of the area the limits of the marine transgressions associated with the Woodston Beds, March Gravels, Abbey Gravels and subsequently the Barroway Drove Beds lie within a zone 5–10 km wide.

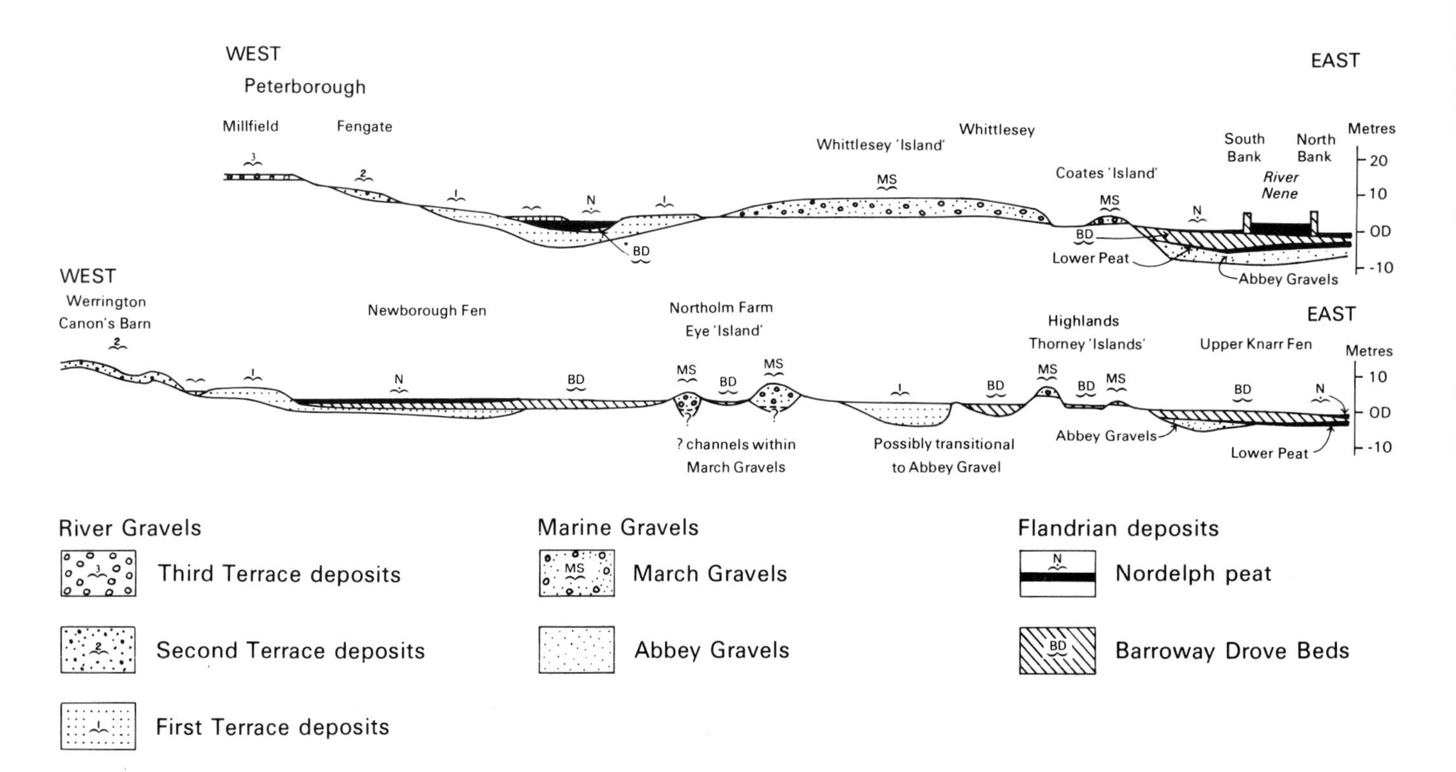

Figure 12 Relative altitudes of the Terrace deposits, marine gravels and Flandrian deposits

SIX

Flandrian deposits

Skertchly (1877) established the stratigraphy of the Flandrian of the Fens, and his work was refined by the studies of Godwin and Clifford (1938), Godwin (1940) and the members of the Fenland Research Committee. The nomenclature was subsequently formalised by Gallois (1979) who defined broad lithological units bounded by marine transgressions and regressions in the southern part of The Wash, north-east of the Peterborough district. The three sets of names are equated in Table 3. The ages of the Terrington Beds and Barroway Drove Beds are inferred.

The Flandrian was a period of aggradation with minor interruptions, a process which continues today with the accumulation of sand flats and salt marshes at the edge of The Wash.

Table 4 Stratigraphic nomenclature of the Flandrian deposits

Approximate age in years BP	Gallois (1979)	Godwin (1940)	Skertchly (1877)
present to at least 2250	Terrington Beds	Upper Silt	Warp
present to at least 3250	Nordelph Peat	Upper Peat	Peat
?2250 to about 7500	Barroway Drove Beds	Fen Clay	Buttery Clay
3390 to at least 7690	Lower Peat	Lower Peat	Lower Peat

Skertchly grouped the Warp and Buttery Clay as Fen Silt.

LOWER PEAT

The Lower Peat rests on the Crowland Bed and is generally overlapped by the Barroway Drove Beds. It is thus only rarely exposed in shallow excavations and is known in the Peterborough district mainly from boreholes. To the north and north-west of Crowland [240 100] it is present as a continuous sheet, less than 1 m thick and dipping gently to the north-east. Locally it may split into two thin leaves, as in a borehole [2373 1527] north-west of Gull House. The Peat is absent along a tract north-east from Crowland to Moulton East Fen on the northern margin of the district. To the south-east there is a patchy distribution around the 'islands' of Eye and Thorney which may indicate erosion of the peat before the deposition of the Barroway Drove Beds (Figure 13).

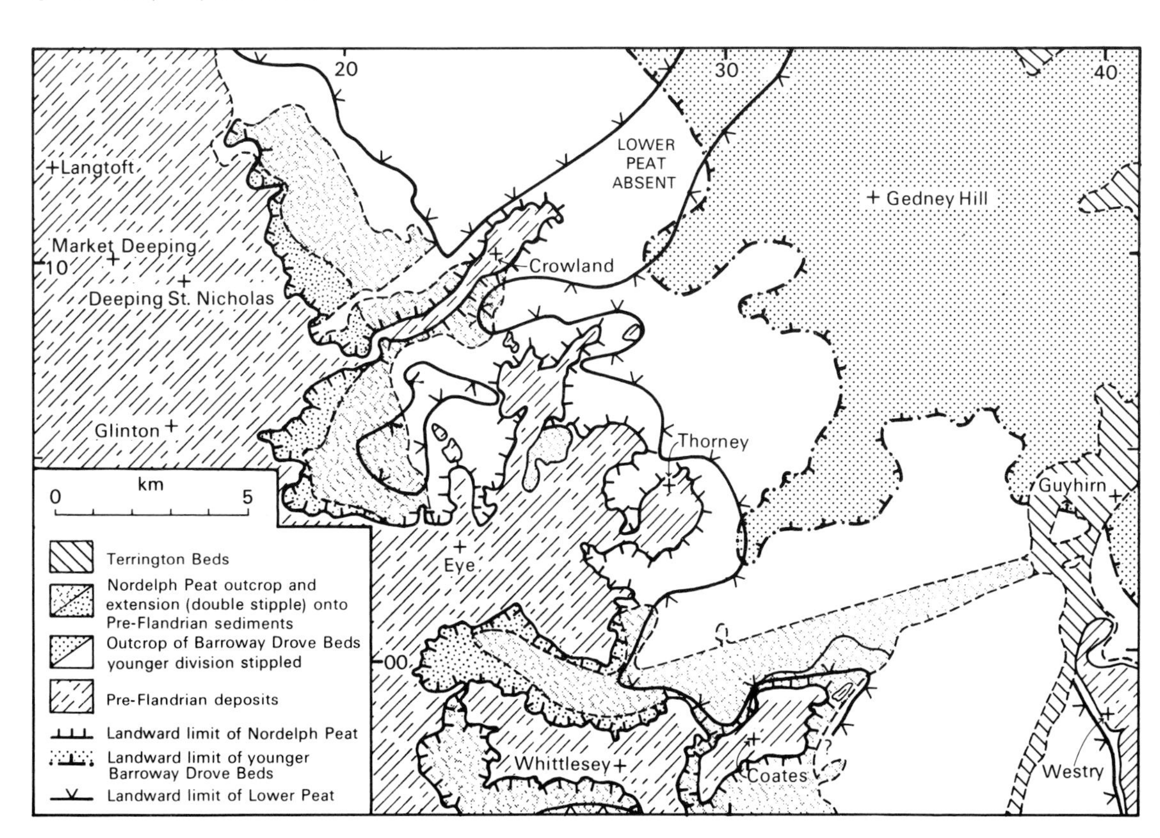

Figure 13 Western limit of Lower Peat and Barroway Drove Beds

East of Stanground [TL 210 970] the First Terrace and Abbey Gravels fill old channels of the Nene which run to the north and south of the Whittlesey 'island'. These deposits are themselves cut by later channels which are floored by the Lower Peat. In the southern channel the peat descends from about 1 m above OD near Stanground to about 5 m below OD south of Whittlesey. It is thicker in this area than to the north, 2.7 m being recorded near Manor Farm [TL 2708 9604]. Lower Peat is present in the northern channel at Stanground North, below Flag Fen [230 000], beneath The Wash [TL 280 990] and the Knarr fens [310 020]. A section [TL 2702 9972] near the Dog-in-a-Doublet, north of Whittlesey, showed thin (0 to 10 cm) Lower Peat upon the gravels of the First Terrace and almost joining with the overlying Nordelph Peat as the intervening Barroway Drove Beds thin to the west (Figure 14). The Peat extends north-eastwards in the buried channel from the Whittlesey area to Guyhirn [3950 0332] where it lies at 9.75 m below OD. At this locality there is an early depression filled by till which was channelled prior to the deposition of the Abbey Gravels, and which themselves were channelled before the accumulation of the Peat. Between Whittlesey and Guyhirn the Lower Peat splits into two leaves which are separated by up to 1.2 m of Barroway Drove Beds. To the north of Guyhirn it appears that the Peat is absent along the eastern margin of the district, near Tholomas Drove [400 064].

The Lower Peat contains tree stumps in situ, and the pollen content indicates an upward change from oak through pine forest to a sphagnum moss bog. This paralleled a change from alkaline to acid groundwater which followed a rising water table. The Lower Peat is deepest in the east of the district and ascends to the west and south-west, co-incidentally with diachronous decrease in age from at least 7500 years in the east to about 3400 years in the south-west. It is probable therefore that the location of peat development moved gradually inland with time.

BARROWAY DROVE BEDS

The Barroway Drove Beds crop out over the greater part of the Fens and consist mainly of dark grey, very soft, slightly humic and silty clays. They commonly contain traces of black carbonaceous rootlets and in places stems of the reed *Phragmites* which grew as the clays were deposited. The clays are cut by a complex network of channels filled by silt and fine sand, individual channels ranging in size from less than 1 m deep and 2 m across to 10 m deep and up to 1 km wide. The channel sediments are dark grey at depth but pale brown when oxidised at outcrop. This colour contrasts with the dark grey of the surrounding clay soils and it has been possible to map out the distribution of the channels from aerial photographs (Plates 3 and 4). Figures 3 and 4 in the margin of the geological map show this pattern but only the major channels are shown on the face of the map. As well as the contrast in colour, the silt and sand of the channels has compacted less than the clay and peat of the surrounding ground. This has resulted in the channels now forming low ridges, generally less than 1 m high and known locally as roddons. Figure 15 illustrates the relationship between two sets of roddons in the lower and upper parts of the Barroway Drove Beds. There were apparently two stages of channel development, one prior to the growth of the lower leaf of the Nordelph Peat and another prior to the accumulation of the upper leaf.

The maximum thickness recorded in the district was in a borehole at Poplartree Farm [4056 1343] where a 11.5 m sequence consisted of 9.25 m of silty clay on 1.25 m of sand

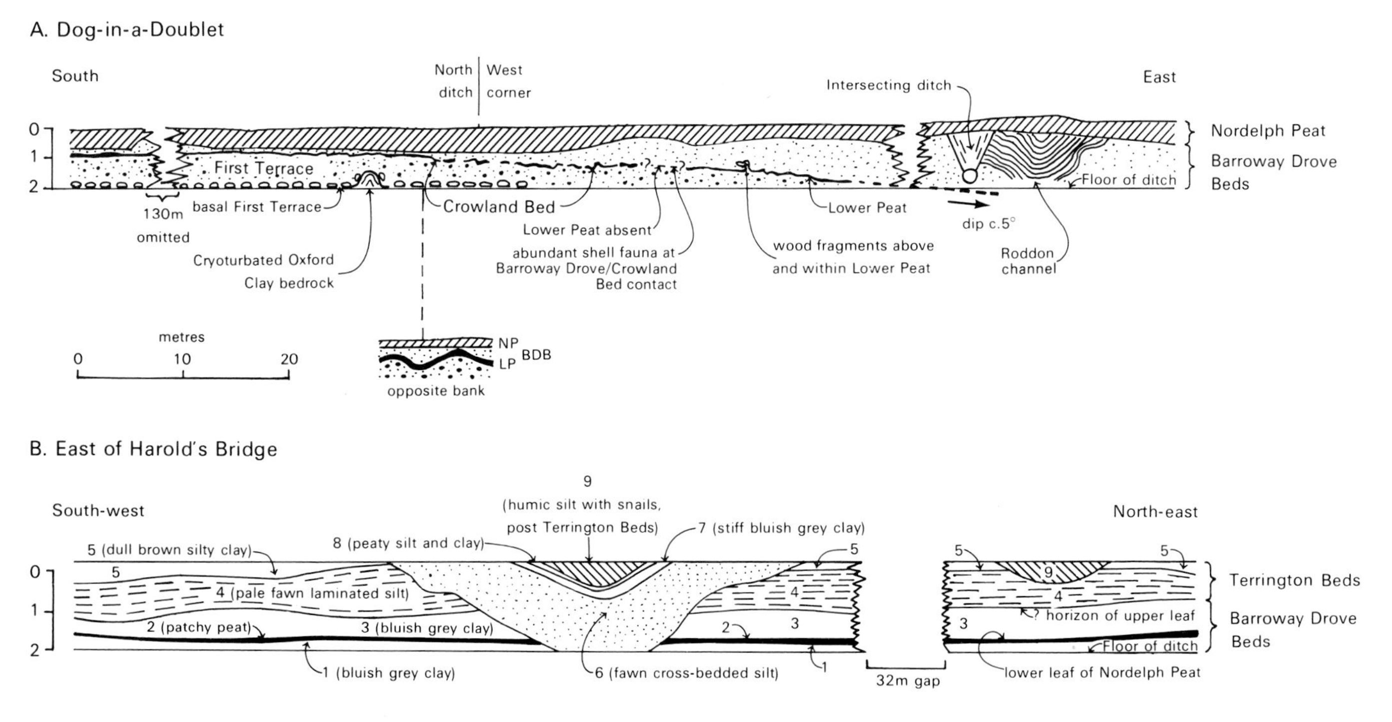

Figure 14 Section through Flandrian deposits at Dog-in-a-Doublet and Harold's Bridge

Plate 3 Aerial photograph of the Barroway Drove Beds and Nordelph Peat at Hop Pole, north-east of Market Deeping [TF 186 138].

The major creeks (roddons) are clearly visible as pale ribbons of silt meandering across the clay outcrop. The final stage of infill of the creek is marked by a dark narrow outcrop of humic silt. The Nordelph Peat obscures the pattern except for the major roddens in the west where the peat destruction has been more rapid over these raised features than in the adjacent areas.

with a basal shell bed. The base rested on the Crowland Bed at about 10.2 m below OD.

The base of the Barroway Drove Beds is diachronous like the underlying Lower Peat, rising gently towards the Fen margin in the south-west, and becoming progressively younger with the rising Flandrian sea level. Near the Fen edge it may overlap the Peat to rest upon older drift or solid rocks. The estuarine and salt-marsh deposition of the Barroway Drove Beds was contemporaneous with the development of peat in the more landward areas but the dividing line moved to the north-east or the south-west with the retreat or advance of the sea. At times of lower sea level the peat extended seawards and some peat horizons are laterally persistent whilst others are localised. Figure 2, in the margin of the geological map, shows six peat horizons within the Flandrian south-west of Guyhirn.

Plate 4 Aerial photograph of the Barroway Drove Beds north-east of Thorney near Priests Farm [TF 299 063].

This location was more seaward than that shown in Plate 3. The creek drainage system is more complex and includes very large roddon units. The final infill stage is visible in only a few examples. The effect of crop cover is to obscure the structure in several areas.
Copyright. The Committee for Aerial Photography, Univesity of Cambridge.

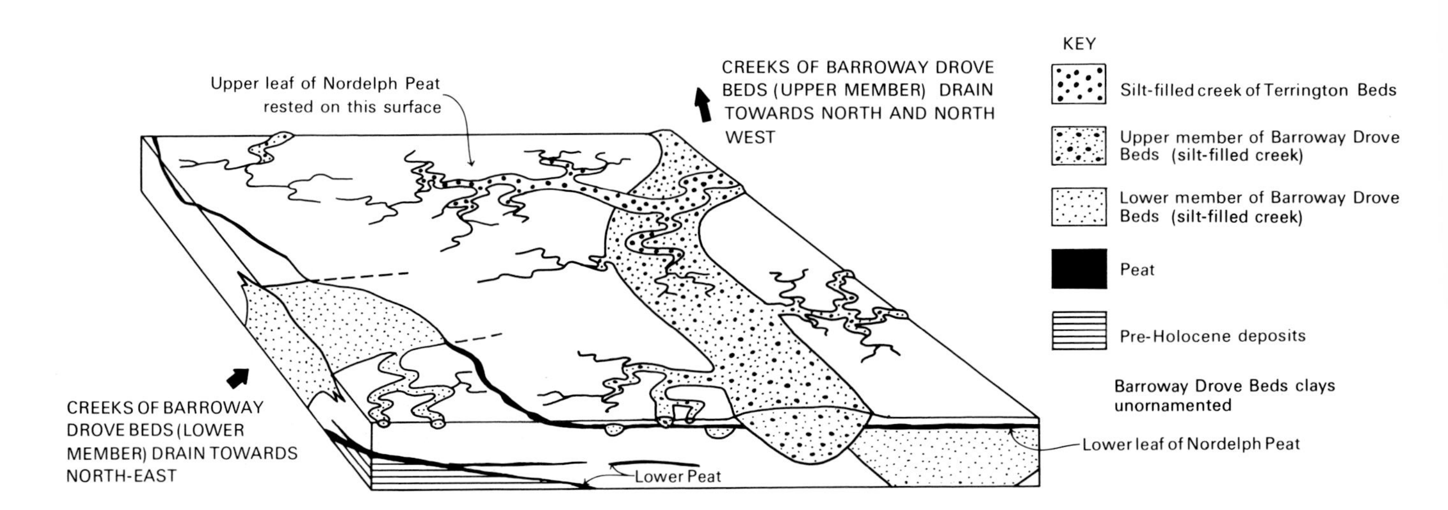

Figure 15 Diagram illustrating the relationship between the members of the Flandrian sequence

The lower leaf of the Nordelph Peat is a thin but persistent horizon which separates the upper and lower parts of the Barroway Drove Beds. It accumulated during a major retreat of the sea which ended the deposition of the lower Barroway Drove Beds. The excellent description by Evans and Mostyn (1979) of the sections in a gas-pipe trench across the eastern part of the district, shows that the lower Nordelph Peat truncates the roddons of the lower Beds (*see* also Figures 1 and 4 in the margin of the geological map). The peat is itself channelled by the roddons of the upper part of the Barroway Drove Beds, these younger channels having different trends from those below (Figure 15).

The upper part of the Barroway Drove Beds is more silty than the lower and may represent more open intertidal flats rather than salt marshes and creeks. A marine regression permitting the accumulation of the upper leaf of the Nordelph Peat ended the deposition of the Barroway Drove Beds.

Marine molluscs, particularly *Cerastoderma edule* and *Macoma balthica*, occur sparsely throughout the formation but are most common in the basal parts of the thicker sequences. MacFadyen (1933, 1938) recorded the presence of foraminifera, and Godwin and Clifford (1938, p.341) described ostracods and diatoms, these being thought to indicate deposition in an area of brackish water penetrated by high tides. MacFadyen (1970) showed that the foraminifera included marine genera mixed with forms more tolerant of brackish or estuarine environments. The vertical variations in the fauna are taken to reflect changes in salinity during deposition.

NORDELPH PEAT

The Nordelph Peat extended beyond the landward limit of the Lower Peat and now crops out mainly in the western part of the Fens. In the Welland basin there is an extensive outcrop from Hop Pole [185 137] southwards through Newborough [203 060] to the northern margins of Peterborough, this outcrop being divided by the artificially banked Welland Wash. Here the Nordelph Peat overlaps the Barroway Drove Beds to rest upon alluvium and First Terrace deposits.

In the north-east part of the district the Peat is split into two leaves, for example at Bone's Gate [400 120], (*see* Figure 1 in margin of geological map), but it occurs as a single bed elsewhere.

In the Nene basin to the south the Nordelph Peat surrounds the March Gravel 'islands' of Whittlesey and Coates and extends eastwards towards Guyhirn as a promontory between the artificial banks of the canalised Nene and Morton's Leam. In this area the Peat overlaps the Barroway Drove Beds, alluvium and First Terrace to rest upon the March Gravels and Oxford Clay. As the lower part of the Barroway Drove Beds thins towards the Fen margin, the Nordelph Peat may become separated from the Lower Peat by only a few centimetres of sediment (Figure 14). Outside of the district to the north and south the two peats come together as a single bed, which in these areas represents unbroken accumulation of peat. The outcrop of Nordelph Peat in the promontory south-west of Guyhirn is limited by the outcrop of the overlying Terrington Beds and these sediments also overlie the two leaves of the Peat in the area of Bone's Gate (geological map, Figure 1).

The outcrop of the Nordelph Peat would originally have been much more widespread, extending eastwards to the outcrop of the Terrington Beds and westwards over parts of the First Terrace outcrops. During the last geological survey areas with more than 0.3 m of peat or peaty soil were included in the peat outcrop. Allowing that Skertchly may have used different criteria, comparison with his boundary (1877, plate 1) shows that peat has been lost over a zone extending up to 4 km to the west during the last century. The loss of peat can be attributed primarily to the effects of drainage which has lowered the water table and facilitated oxidation of the peat. This has encouraged a change from pastoral to arable farming with a resultant deflation of the peat. The Fens are now significantly affected by dust storms, most commonly in the Spring, caused by wind erosion of the ploughed soil. Evidence of the former extent of the Nordelph Peat is now provided only by pockets of humic soil, by peat remnants around and beneath farm buildings and below artificial alluvial tracts such as the Nene and Welland washes and, in places, the Terrington Beds.

Table 4 shows the range of radiocarbon ages of samples taken from the peats.

TERRINGTON BEDS

The Terrington Beds crop out only in the eastern and northern parts of the district, and are named from the King's Lynn district to the north-east (Gallois, 1979). Two facies have been recognised, the first comprising pale brown silty clays and silts which are thought to be the deposits of salt marshes and tidal flats. These occur west of Grangehill Farm [384 151] and near Bone's Gate [402 119]. The second facies consists of pale brown silt and fine sand which fills channels cut into the sediments of the first type and older formations (*see* Figure 3 in the margin of the geological map). Within the Terrington Beds outcrop the lack of contrast between channel fill and surrounding sediments makes it much more difficult to map out the channel system from aerial photography. The lithological similarity between the two facies also results in very little differential compaction of the sediments so that the roddons within the Terrington Beds are more subdued topographically.

Channels of Terrington Beds age extend westward into the outcrop of the Barroway Drove Beds where they may form distinct roddons or occur as a secondary fill in older channels. Within the district the most important channel fill is marked by the outcrop which extends north from Whittlesey Road Farm [TL 377 967] to Guyhirn [403 040].

The Terrington Beds rest disconformably on the upper leaf of the Nordelph Peat and were deposited during the last major transgression of the Flandrian sea which only just reached the district. Salway (1970) considered that deposition started at some time between 1300 and 300 BC and ceased before the Roman occupation of the silt Fens (the Terrington Beds outcrop) which he considered to be not earlier than about 80 AD. At the same time, however, the open creeks would have carried sea water and marine sediment farther south-west than the shoreline. There is evidence that a major flood carried sediment as far inland as the Car Dyke, a Roman drainage channel which runs north-west from Peterborough. Shell material from a borehole, in a sand-filled channel in the Terrington Beds of the Spalding district to the north, gave radiocarbon ages ranging from about 1550 years to about 1900 years BP.

ALLUVIUM

The flood plain deposits of the Nene and Welland consist of clay and silt with small scattered pebbles, locally with a layer of gravel at the base. The alluvium fills a channel cut into the gravels of the First Terrace but overlaps it as a thin veneer to cover the whole of the flood plain. The clays and silts are dark grey and humic when fresh (Plate 5) but they weather brownish grey with a cuboidal fracture, similar to that of the clays of the Barroway Drove Beds. There are some shelly horizons locally and beds of peat occur within the sequence.

The alluvial channels descend to the east. West of Longthorpe a borehole [TL 1481 9818] proved the base of the channel at 0.4 m above OD and this has descended to 2.4 m below OD at the main line railway bridge at Peterborough [TL 1902 9814] (Horton and others, 1974, fig.9).

A borehole sample of peat [TL 1812 9844], from 0.25 m above the base of the Nene alluvium in Peterborough, yielded wood fragments which gave a radiocarbon age of 3475 ± 100 years BP. Another hole in the same area [TL 1823 9827] provided a peat sample from 0.55 m above the base of the alluvium with an age of 4460 ± 105 years BP. These results indicate that the channel wandered laterally within the confines of the valley sides and locally eroded earlier alluvial sediments.

Farther downstream, at the old Peterborough East Station [TL 1943 9788], a borehole proved the alluvium at 4.87 m thick with three thin beds of peat. The sequence here is transitional to that of the Fens and the alluvial clays probably pass laterally into the Barroway Drove Beds.

The geological map shows a network of alluvial channels in the Welland Basin, to the east of Market Deeping and Maxey, while there is an artificial extension of alluvium north-eastwards past Crowland, confined by the drainage banks of the Welland washes. The alluvial deposits are up to 1.8 m thick around Northborough [1626 0886], this being more than 1 m less than that recorded in the topographically confined and hence laterally restricted Welland valley west of the district. In the area of the Deepings the alluvium encloses areas of terrace gravels and probably passes eastwards transitionally into the Fen sequence.

FLANDRIAN HISTORY

About 10 000 years ago, at the beginning of the Flandrian period, the Fenland area was dry land with a relatively low sea level of about 30 m below that of the present, resulting from the Devensian glacial period. From this early stage the Flandrian has been characterised by a progressive but discontinuous rise in sea level, the effects of which have been modified by vertical tectonic crustal movements and by changes in tidal range and coastal geomorphology. These factors have been discussed in detail by authors such as Tooley, Shennan, Heyworth and Kidson. The first two of these (Tooley, 1982; Shennan, 1982), considered that the movement of relative sea level was spasmodic, with rises and falls causing distinct periods of marine transgression and

Plate 5 Section [TL 1535 9734] in alluvial deposits of the River Nene north of Ham Farm, Orton Waterville. The crumbly-textured humic alluvial clay rests upon gravels of the First Terrace. (A 11636).

regression, which resulted in the onlap or offlap of the sediments. Shennan (1982, p.59) defined periods of rising and falling sea level chronologically by means of radiocarbon dating of peaty horizons. Other authors such as Heyworth and Kidson (1982) accepted that the rise in sea level was modified by oscillations but considered that these were less than the uncertainties in defining the contemporaneous sea level. These latter included difficulties in defining the levels of peats, particularly in boreholes; the problem in recognising and dating the precise horizon which marks the change in movement of sea level; relating lithological changes to the water table; estimating the effects of consolidation of sediments; and finally and not least, errors in radiocarbon dating. Certainly the trench records of Evans and Mostyn (1979) showed that the level of a peat bed may fluctuate considerably, by as much as 1 m in height over a horizontal distance of 100 m. This is a major variation, considering that the local Flandrian sequences are only of the order of 10 m thick.

Because of the above uncertainties the present account does not describe the Flandrian sediments on the basis of chronological units but in terms of lithostratigraphic divisions. It is considered that peat grew throughout most of the Flandrian, being replaced laterally at intervals by marine sediments. Table 4 shows the radiocarbon ages of samples taken from the peats of the Peterborough district. It lists the boreholes and exposures from which samples were taken and gives their grid references.

Figure 16 is based on Table 4 and is intended to illustrate diagrammatically the different ages of peat accumulation across the district. It should be noted that the vertical scale at each locality in the diagram is one of time rather than depth. Grid references and the experimental error in ages are listed in Table 4 and are omitted from the following account for easier reading.

The oldest peat in the district is the Lower Peat at about 7690 years at Elm Tree Farm in the extreme north-east. To the south at Adventurers' Land peat growth did not start until about 6575 years ago and continued for about 300 years. Following an interruption of about 700 years, peat growth restarted at about 5580 years ago (Shennan, 1982, p.57). Both of these horizons are grouped as the Lower Peat.

At Plash Farm, between Elm Tree Farm and Adventurers' Land, peat growth did not start until about 6080 years ago, but this was perhaps due to the existence of a creek in this area which would have prevented plant growth. Westwards from Adventurers' Land the Lower Peat becomes progressively younger, ranging from about 5140 to 5000 years old in the Welland Wash No. 4 Borehole to only 3390 years near Newborough.

The major pre-Flandrian channels, for example those of the Nene north and south of Whittlesey, and that of the Welland, were not totally filled by peat before the marine incursion of the Barroway Drove Beds. There may also have been some erosion of the Lower Peat prior to the deposition of the Barroway Drove Beds, for example in a depression between the March Gravels north-east of Eye and the First Terrace/Abbey Gravels south-west of Crowland. This channel descends to 3 m below OD and is locally cut into Oxford Clay. It may be an old channel of the Nene.

Table 4 Radiocarbon ages from the Flandrian of the Peterborough district

Nordelph Peat	Middle Peat	Lower Peat
1845 ± 50 AL		
2220 ± 50 Ne		
2270 ± 50 PaF		
2510 ± 50 PF		
2550 ± 60 WW5		
3050 ± 50 PaF		
3080 ± 200 GW		
3250 ± 50 GH		
		3390 ± 40 Ne
River Nene Alluvium		
3475 ± 100 P1		
	3860 ± 80 WW4	
	4030 ± 80 WW4	
	4180 ± 75 AL	
	4310 ± 140 SoF	
	4340 ± 60 GW	
4460 ± 100 P2	4460 ± 80 SF	
	4500 ± 50 AL	
	4520 ± 70 PF	5000 ± 70 WW4
		5140 ± 60 WW4
		5580 ± 70 AL
		6010 ± 200 SF
		6080 ± 60 PF
		6275 ± 125 AL
		6415 ± 185 AL
		6575 ± 95 AL
		7690 ± 400 EF

Abbreviations		Grid reference
AL	Adventurers' Land	3567 0182
EF	Elm Tree Farm	4010 1487
GH	Gedney Hill	3344 1084
GW	Guyhirn Washes	3810 0198
P1	Peterborough	TL 1812 9844
P2	Peterborough	TL 1823 9827
PF	Plash Farm	3873 0530
PaF	Park Farm	4018 1608
SF	Sycamore Farm	3370 1113
SoF	South Farm	3024 0210
Ne	Newborough	1953 0524
WW4	Welland Wash 4	2459 1432
WW5	Welland Wash 5	1923 0824

Dates are given in years before present (BP).
The Newborough site was originally listed as The Firs, Werrington

Following the deposition of the lower part of the Barroway Drove Beds there was a significant emergent phase which is marked by the growth of the 'Middle' Peat which has been recorded between the Welland Washes and Guyhirn. Shennan dated this period at about 4500 to 4180 years at Adventurers' Land and this compares with similar ages at South Farm, Guyhirn Wash and Plash Farm, although at Welland Wash No. 4 Borehole to the north-west, the peat is younger, from about 4030 to 3860 years. South-westwards towards the Fen margin the 'Middle' Peat and the Lower Peat are overlapped by the Barroway Drove Beds but peat growth was probably continuous in some places on the landward margin of the Barroway Drove Beds sea. At its maximum this shoreline extended from the vicinity of Deeping Common [160 120] to Newborough [200 060], Eye Green

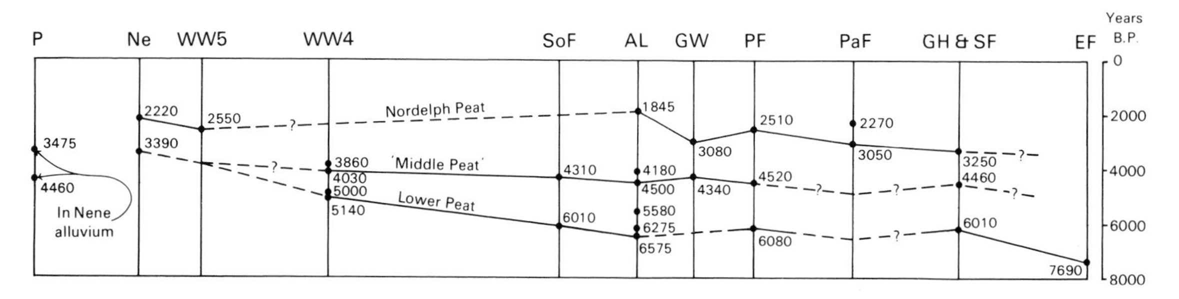

Figure 16 Ages and variations in Flandrian peat Horizons

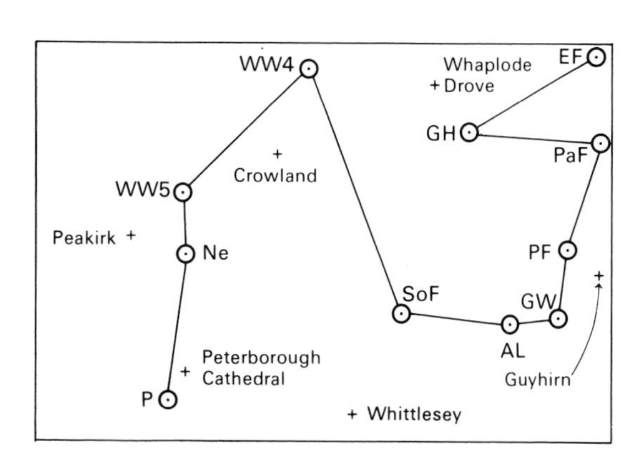

[230 040] and around the 'island' north-east of Eye to Thorney. South of here a major embayment extended almost to Peterborough with isolated islands at Whittlesey, Coates and Westry. The Lower Peat (3390 years) and the Nordelph Peat (2220 years) at Newborough are separated by only 1 m of Barroway Drove Beds, suggesting that the maximum of the marine transgression occurred at about 3000 years ago.

Following this maximum a regression permitted the development of the Nordelph Peat. It might be expected that with the progressive retreat of the sea, the re-establishment of peat growth would become younger towards the north-east. This simple pattern is, however, probably complicated by the geography of the time and peat growth may have been inhibited in some areas by the existence of marine creeks or embayments. At Guyhirn the base of the Nordelph Peat is dated at about 3080 years, but the Peat splits into two leaves to the north, and the lower of these is dated at about 3050 years at Plash Farm and at about 3250 years at Gedney Hill. In contrast, at Adventurers' Land west of Guyhirn, growth of the Nordelph Peat did not commence until about 1845 years ago (Shennan, 1982, fig.3).

In the seaward part of the Fenland, peat growth was stopped by the last major marine incursion which deposited the Terrington Beds. This however only just reached the eastern part of the Peterborough district.

MODERN DRAINAGE SYSTEM

Although the positions of the early Flandrian and Pleistocene drainage channels of the district are known approximately from borehole data, there is no clear picture of the channels which flowed seawards at the time of the Nordelph Peat. At Peterborough, the existence of two channels of the Nene beneath the alluvium indicates that the river continued to flow in two branches in historic times. The Nordelph Peat is probably equivalent to the uppermost part of the alluvium. The northern alluvial tract of the Nene narrows across the Nordelph Peat outcrop of the Flag Fen [230 000], along the line of Cat's Water, suggesting that the flood waters were dispersed throughout the peat-covered fens. Although much of the peat has now been removed from this area, the line of Cat's Water remains and it is possible that this branch of the Nene had a channel within the Nordelph Peat at that time. Skertchly (1877) considered that this route continued along the line of the Old South Eau to near Holbeach Drove Common [310 104] and then eastwards along the Old Holland Drain parallel to the winding road to Parson Drove. From here it flowed northwards, probably along the line of what is now called Lady Nunn's Old Eau. The other alluvial tract of the Nene runs southwards around Stanground and then beyond the present district.

These two routes, both of which were channels along the edge of the peat were also suggested by Evans (1979) as post-Roman courses. Skertchly (1877) also proposed two other channels, one north of Whittlesey along the line of Morton's Leam and the other to the south of the 'island', the King's Dyke. There is no evidence of sediment associated with these courses.

The Welland divided into numerous channels as it left the uplands and deposited a complex pattern of patches of alluvium in hollows within the First Terrace gravels. The alluvium can only be traced north-eastwards across the outcrop of the Nordelph Peat along the line of The Wash [195 085] and Crowland High Wash [225 100]. This strip is enclosed between flood banks to accommodate high flood water and has thus continued to receive alluvium up to the present. Up to 1 m of alluvium overlies and preserves the Nordelph Peat along this strip. Skertchly (1877, p.67) suggested that the northern branch of the Nene may have had a dual-flow cross connection with the Welland near Crowland and that, before the 17th century, the stream may have taken a large proportion of the waters of the Welland into the South Eau and thence to Wisbech.

Whatever the original courses, the present drainage of the Fenland is almost entirely artificial. The first man-made channels were probably excavated by the Romans. The most important was the Car Dyke, a catch-water channel designed to collect the water from the upland streams and direct it northwards along the margin of the Fens (Skertchly, 1877, p.15). It probably also functioned an as important navigational channel. The Romans avoided the peat areas, developing farms either at the Fen margins or on higher ground such as the roddons within the Crowland [240 100]–Whaplode Drove [320 133]–Parson Drove [370 085]–Guyhirn [400 035] district. In these areas there are several ancient steep banks which were probably built as flood banks or sea walls; their date of origin is uncertain. By Saxon times the sea lay well beyond the present district, the so-called Roman Bank between Spalding and Wisbech probably marking the limit of the sea at that time. The pattern of these banks suggests several periods of enclosure and possibly of flood protection. The most inland bank, the Green Bank, south of Crowland, probably joined a bank along the old South Eau, to Cloughs Cross [368 093]. South from here the important Murrow–Guyhirn Bank formed a major barrier. From North Fen [284 100], the Dowsdale Bank runs towards Whaplode and Astwick Grange. At Cloughs Cross, Marshall's Bank ran east-north-eastwards and then northwards to bifurcate north at Harold's Bridge [387 113] into the Elloe and Treading banks. At times, storms broke through the banks to leave gaps or gulls, some of which remain today as indented stretches of bank, for example at Guyhirn Gull [389 043]. The construction of these barriers necessitated changes in the drainage system and continued the work of the Romans, who in addition to the Car Dyke, probably dug other drains extending out into the Fens.

In the south, Bishop Morton is reputed to have constructed Morton's Leam (1478–1490), a canalised course of the northern branch of the Nene from Peterborough to Guyhirn (Skertchly, 1877, pp.45–69, 73).

The greatest drainage development came in the 17th century, following the passage of an act during the reign of Elizabeth I to drain all the low-lying parts of England (Skertchly, 1877, p.34). The major works started after the death of the Queen. Vermuyden straightened the northern course of the Nene in 1609 but his major works in the Fens belong to the period 1618–1637. The 4th Duke of Bedford in conjunction with Vermuyden organised a group of financiers to fund the drainage of the Bedford Level, a low-lying area stretching from Crowland to Ely, which included most of the Fenland of the Peterborough district. This group was known as the Adventurers and their name survives in Adventurers' Land, west of Guyhirn. At that time Bevill's Leam was cut and now forms part of the Twenty Foot River south-west of Guyhirn; Morton's Leam and the Peakirk drain from Peakirk [170 067] Cloughs Cross and Hill's Cut near Peterborough were also improved. Possibly at this time the area [320 020] then known as Knarr Lake, south-east of Thorney, may have been drained. Work continued from 1649–51 with the embanking (1726) of the River Nene along its present course from Stanground to Guyhirn and the extension of the Twenty Foot River. Although other drains have been cut since and the directions of flow of dykes changed, the earlier channels still form the basis of the modern drainage system.

SEVEN

Structure

Little is known about the structure of the sub-Mesozoic rocks of the Peterborough district, but the Jurassic rocks which cover the district have a very simple structure with a gentle regional dip to the east or the east-south-east (Figure 7). This is modified by several faults, the most important being the Marholm Fault which trends approximately east–west just north of Marholm [150 024]. This throws about 20 m down to the north bringing Oxford Clay on the north side against Upper Estuarine 'Series', Blisworth Limestone and Clay, Cornbrash and Kellaways Beds on the south. There are several faults subsidiary to the main one, for example near Cat's Water Farm [247 040] where there is a partly fault-bounded outcrop of West Walton Beds. In south-west Peterborough a fault trending roughly north–south has been identified from borehole data. In the north-west of the district near Langtoft [123 124], a fault trending north-north-west to south-south-east and throwing down to the west has been deduced from borehole evidence showing displacement of the Kellaways Sand.

There is some evidence for the presence of faults beneath the extensive drift cover. Figure 7 shows two such faults which affect the base of the Lincolnshire Limestone. The more southerly runs through Etton [140 065] and is parallel to the Marholm Fault, also throwing down to the north. The more northerly runs north-east from Crowland [240 100] and throws down to the north-west.

EIGHT

Economic geology

The Peterborough district is predominantly rural with agriculture as the major industry. The extraction of sand and gravel for building and the construction industry continues on a large scale in some parts while the Peterborough area is well known for the manufacture of bricks, the material for which has been obtained from the Oxford Clay.

SAND AND GRAVEL

All the terrace and marine gravel deposits have been exploited. The outcrops contain many small pits which have been dug for local or farm use. Workings in the First Terrace are the most widespread and these deposits contain the main resource for the future. There are large areas of quarrying in the north-west at Langtoft Fen [135 145], west and south of Maxey [130 080], at Tanholt Farm [235 016] and at Northey [TL 240 986]. In the past, gravel was dug extensively [210 988] in the Fengate area of south-east Peterborough, north-west of Market Deeping [112 110] and north-east of Peakirk [180 080].

The March Gravels have been exploited north of Coates [TL 308 982] and west of Eldernell [TL 313 988], on the Whittlesey 'island' south of Bassenhally Farm [TL 289 982], east of the town [TL 285 967] and at Lattersey Hill [TL 285 961]. West of the town the deposit forms an overburden in the Oxford Clay pits and is dug prior to the exploitation of the brick clay. Deposits of the Second Terrace were extensively exploited at Woodston [TL 181 974, 185 973 and ?188 973] and near Oxney [22 008]. Deposits of the Third Terrace were dug in pits at Orton Waterville [TL 160 961, 160 964, 162 963]. The economic potential of the sand and gravel was assessed in detail by Booth (1981, 1982 and 1983). The various deposits are broadly similar in composition since they have all been derived from the Jurassic limestones and the overlying glacial drifts. This is illustrated by the dominance of limestone and flint in the pebble counts of the gravels shown in Table 5 which was prepared from data in the published Mineral Assessment Reports.

The deposits of the First Terrace and Abbey Gravels are the most important economically. The lithology varies from very clayey, sandy gravel to gravel, but most of the deposits are sandy gravel. The composition varies within the following ranges: gravel 21–44 per cent, sand 50–67 per cent and fines 3–22 per cent. The resource assessment boreholes proved thicknesses of up to 7.6 m but the average thickness is of the order of 3 m.

The Second Terrace and March Gravels comprise gravel and clayey, sandy gravel with a range in composition as follows: gravel 32–47 per cent, sand 47–56 per cent and fines 6–12 per cent. The limited amount of information suggests that the thickness may range up to 7.7 m but that the average is 1.6 to 2.0 m.

BRICK CLAY

The Oxford Clay is the major source of brickmaking clay in Great Britain, some 33 per cent of total national production using this as a raw material. The Peterborough region includes the second most important complex of brickworks after the Bedford area, the London Brick Company being the only producer in both areas.

The Peterborough region has six operational works of which the two most recently developed, Kings Dyke (1969–1974) [TL 245 975] and Saxon (1972) [TL 255 972],

Table 5 Mean composition by weight of the gravel fractions of the gravel deposits of the Peterborough district

Formation	Size of gravel fraction	Limestone including chalk	Flint	Ironstone	Sandstone	Quartzite	Others
First Terrace /Abbey Gravels[1]	+ 8 – 16 mm	29	33	8	20	10	0
First Terrace /Abbey Gravels[2]	+ 4 – 64 mm	33	42	8	5	11	1
Second Terrace /March Gravels	+ 4 – 64 mm	32	48	7	3	9	1

1 In the area of Sheet TF 11
2 In the area of Sheets TF 20 and TL 29 (including part of TL 29 not in the Peterborough 158 sheet area)

are situated west of Whittlesey, within the area of the geological map. These two account for 60 per cent of the brick production around Peterborough, the remainder being produced in the brickworks at Dogsthorpe [207 023], north-east of the city, and at Northam, Orton and Beebys which lie to the south of the district.

Calloman (1968) listed the characters on which the pre-eminence of the Oxford Clay as a brick clay was established. These are 'the low water content, the presence of free lime, the clay mineral composition, the organic content and the absence of impurities'. The Fletton process was developed in the long-disused brickworks south of Old Fletton and involves the semi-dry pressing of bricks. The weathered material, the callow, usually 5 m thick, is first removed to reveal fresh Oxford Clay, or knotts. The latter has a low (20 per cent) water content and can thus be ground, as extracted, to particles which are suffcently small to be compressed and moulded into bricks with sufficient strength to be stacked directly in the firing kiln without a drying stage. The mineral character of the clay determines the way in which the brick burns and its strength.

The Lower Oxford Clay conforms to the ideal characteristics of a brick clay, having a lime content of between 5 and 15 per cent and a composition dominated by clay mica minerals (illite types in this case). It contains only a very few impurities such as thick-shelled calcitic fossils and small calcareous nodules. The Lower Oxford Clay fossils are generally preserved with thin fragile shells which readily break up during grinding. Any massive septarian limestone nodules can be easily removed during excavation. Possibly the most important factor is the highly bituminous character of the shale beds of the Lower Oxford Clay. The organic content is sufficient to reduce the additional fuel requirement to between one third and one sixth of that necessary for the firing of other brick clays. These optimum characters occur in combination only in the beds of the Lower Oxford Clay, so the brick clay is extracted from this division. Very small quantities of the underlying Kellaways Beds, the basal strata of the Lower Oxford Clay or the overlying Middle Oxford Clay can be incorporated but would destroy the process if present in larger quantities. The process and the raw material produce an economical brick of such popularity that the name of the village of Fletton has become synonomous with the standard or common brick.

The Lower Oxford Clay has very limited outcrops in the district, but is present beneath the drift deposits along the entire marginal zone of the Fens. It is extremely uniform in character and is a suitable brick clay throughout this area. The availability of material is therefore not limited by geological factors but by amenity and environmental considerations.

In the future, worked out pits will be filled and restored. Already the pits immediately south of Old Fletton have been filled with the waste pulverised fuel ash from power stations in the Trent Valley and the reclaimed land has been cultivated for ten years.

LIMESTONE AGGREGATE, BUILDING STONE AND LIME

The Lincolnshire Limestone, Blisworth Limestone and the Cornbrash were formerly quarried on a small scale in response to local demand. The Cornbrash, although thin, weathers to a rubble of small blocks that were easily exploited for aggregate and rough road material. The most important source of material, the Lincolnshire Limestone, was exploited in one large quarry [123 028] west of Marholm. This has been abandoned though the rock is still worked immediately to the west of the district. It was used for building stone and as a source of burnt lime, but now only provides aggregate for fill for roads and buildings.

WATER SUPPLY

The Lincolnshire Limestone is the only formation in the district that is sufficiently permeable to allow exploitation as an aquifer. It is restricted to the west and north of the district (Figure 7) where it is mostly overlain by younger deposits, cropping out in small areas near Castor and Marholm. In the past, it provided the water supply for Peterborough but, with the expansion of the population and associated industry, water is now also obtained from Rutland Water, near Empingham to the west of the district. This is a pumped storage reservoir with a capacity of 124 million m^3. At times of high river flow, water is extracted from the River Nene at Wansford and the River Welland at Tinwell and pumped to Rutland Water. Additional supplies are obtained from streams that drain into the reservoir.

Despite this more recent use of surface water, the Lincolnshire Limestone remains an important source of water in and to the north and north-west of Peterborough, providing supplies from many boreholes and wells of which the major source is the Etton Pumping Station of the Anglian Water Authority. The water resources in the limestone are recharged by rain falling on the outcrop of the aquifer west and north-west of the district. The average annual rainfall on the outcrop is about 600 mm, but about two-thirds evaporates and only 200 mm reaches the water table, principally in the six winter months. In addition to replenishment by rainfall, the limestone is also recharged by seepage from rivers flowing over the outcrop and by surface run-off discharging into swallow holes.

The Lincolnshire Limestone is extremely permeable. Artesian flows of more than 22 000 m^3/d have been recorded in south Lincolnshire where some of the largest yields in Britain are obtained from individual wells. The pumping station at Etton consists of four wells and the total yield is about 16 000 m^3/d. The aquifer owes its high permeability to a network of fissures which transports the water to wells. The fissures are partly replenished by water stored in microfissures and pores in the matrix of the rock. The transmissivity can be as high as 5000 m^2/d, and locally even 10 000 m^2/d, allowing rapid water flow of the order of ten metres per day.

The aquifer is overlain by the relatively impermeable Upper Estuarine 'Series' which acts as a confining bed. To the west of the district, where the aquifer crops out, the top of the saturated zone, or the water table, occurs some distance below the surface. But, where confined, the limestone is fully saturated and the water level in wells that penetrate into it rises above its top and in some areas water overflows at the surface. This is because of the water pressure in the unconfined area to the west. At the present time water will overflow naturally from wells over much of the Fenland in the north-western part of the district. Originally the overflowing area was larger but it has been reduced as development of the water resources has lowered water pressure in the aquifer. The rest water level over most of the district is now less than 10 m above sea level although in an outcrop area to the west it exceeds 100 m above sea level.

As the water flows through the limestone from west to east, from the outcrop into the confined area, the chemical quality changes. In the outcrop the ions in solution are predominantly calcium and bicarbonate derived from the reaction of rain water, that has dissolved carbon dioxide in the soil, with the calcium carbonate of the limestone; magnesium and sulphate ions are also important constituents. As a consequence the water is hard. Water of this chemical type persists into the confined area but over much of the confined area the water has been softened by natural ion exchange. The calcium and magnesium ions in the water are replaced by sodium ions from minerals in the aquifer matrix. The result is a soft sodium bicarbonate water. In the central part of the district the non-carbonate hardness is reduced to zero by this reaction and the carbonate hardness is less than 100 mg/l with minimum values of less than 25 mg/l. In the west of the district the carbonate hardness is about 200 mg/l and the non-carbonate hardness 50 to 100 mg/l.

The chloride concentration of the water is about 25 mg/l in the outcrop area but this steadily increases to the east. The maximum recommended chloride concentration for a potable water is 200 mg/l which is attained along a general north-south line extending north from Peterborough. To the east of this line values rise to more than 1000 mg/l as water moving downdip from the outcrop mixes with a relatively stagnant saline water that occurs in the aquifer in the east of the district.

Thus a hard calcium bicarbonate water in the extreme west of the district passes into a soft sodium bicarbonate water in the centre, while in the east the water becomes increasingly saline.

Although water can move rapidly from the outcrop through the fissure system in the limestone, the water in the confined aquifer contains a significant proportion of water that has been stored in the aquifer matrix for many hundreds or thousands of years. The mean age of water pumped from wells in the confined aquifer of the district may be between 10 000 and 20 000 years and thus part of it was precipitated during the Pleistocene period.

The Lincolnshire Limestone is a significant groundwater resource. Abstraction from wells is carefully managed by the Anglian Water Authority to ensure that over-pumping does not take place and that water levels and the quality of the water are maintained. The maximum water requirement forecast for Peterborough and the surrounding area by the early part of the next century, is about 130 000 m^3/d. The yield from the two sources available, the Lincolnshire Limestone and Rutland Water, could be increased if they were used in conjunction. Consequently it has been proposed that groundwater should be pumped whenever it is plentiful but that during droughts greater use should be made of Rutland Water.

REFERENCES

ASHTON, M. 1980. The stratigraphy of the Lincolnshire Limestone Formation (Bajocian) in Lincolnshire and Rutland (Leicestershire). *Proceedings of the Geologists Association, London*, Vol. 91, 203–223.

ASLIN, C. J. 1968. Upper Estuarine Series. 233–237 in *The geology of the East Midlands*. SYLVESTER-BRADLEY, P. C. and FORD, T. D. (editors). 400 pp. (Leicester: University Press.)

BADEN-POWELL, D. F. 1934. On the marine gravels at March, Cambridgeshire. *Geological Magazine*, Vol. 71, 193–219.

BOOTH, S. J. 1981. The sand and gravel resources of the country between Stamford and Peterborough. Description of 1:25 000 sheets TF 00 and TF 10. *Mineral Assessment Report of the Institute of Geological Sciences*, No. 80. 126 pp.

— 1982. The sand and gravel resources of the country around Whittlesey, Cambridgeshire: Description of 1:25 000 sheets TF 20 and TL 29. *Mineral Assessment Report of the Institute of Geological Sciences*, No. 93. 109 pp.

— 1983. The sand and gravel resources of the country between Bourne and Crowland, Lincolnshire. Description of 1:25 000 sheet TF 11 and parts of TF 01 and TF 21. *Mineral Assessment Report of the Institute of Geological Sciences*, No. 130. 191 pp.

BROWN, A. E. 1967. Great Billing. *Bulletin of the Northants Federation of Archaeological Societies*, Vol. 2, 5–6.

CALLOMON, J. H. 1956. The ammonite succession in the Oxford Clay and Kellaways Beds at Kidlington, Oxford and the zones of the Callovian Stage. *Philosophical Transactions of the Royal Society, London*, No. B239, 215–264.

— 1968. The Kellaways Beds and the Oxford Clay. 264–290 in *The geology of the East Midlands*. SYLVESTER-BRADEY, P. C. and FORD, T. D. (editors). 400 pp. (Leicester: Leicester University Press.)

CASTLEDEN, R. 1976. The Floodplain gravels of the River Nene. *Mercian Geologist*, Vol. 6, 33–47.

— 1980. The Second Terraces of the River Nene. *Mercian Geologist*, Vol. 8, 29–46.

COPE, J. C. W., GETTY, T. A., HOWARTH, M. K., MORTON, N. and TORRENS, H. S. 1980a. A correlation of Jurassic rocks in the British Isles. Part 1. Introduction and Lower Jurassic. *Special Report of the Geological Society of London*, No. 14. 73 pp.

— — — — — 1980b. A correlation of Jurassic rocks in the British Isles. Part 2. Middle and Upper Jurassic. *Special Report of the Geological Society of London*, No. 15. 109 pp.

EVANS, R. 1979. The early courses of the River Nene. *Durobrivae*, No. 7, 8–11. (Nene Valley Research Committee, Archaeological Field Centre, Hame Lane, Orton Waterville: Peterborough.)

— and MOSTYN, E. J. 1979. Stratigraphy and soils of a Fenland gas pipeline. Internal Report, ADAS Land Service, Ministry of Agriculture, Fisheries and Food. 26 pp.

FRENCH, C. A. I. 1982. An analysis of the molluscs from an Ipswichian interglacial river channel deposit at Maxey, Cambridgeshire, England. *Geological Magazine*, Vol. 119, 593–598.

GALLOIS, R. W. 1974. The Wash Water Storage Scheme. Report on the geological investigations for the feasibility study. Unpublished report of the Institute of Geological Sciences.

— 1979. Geological investigations for the Wash Water Storage Scheme. *Report of the Institute of Geological Sciences*, No. 78/19. 74 pp.

— and COX, B. M. 1977. The stratigraphy of the Middle and Upper Oxfordian sediments of Fenland. *Proceedings of the Geologists Association, London*, Vol. 88, 207–228.

GODWIN, H. 1940. Studies of the post-glacial history of British vegetation. Parts III and IV. *Philosophical Transactions of the Royal Society, London*, No. B230, 239–303.

— and CLIFFORD, M. H. 1938. Studies of the post-glacial history of British vegetation. Parts I and II. *Philosophical Transactions of the Royal Society, London*, Vol. B229, 323–406.

HEYWORTH, A. and KIDSON, C. 1982. Sea-level changes in southwest England and Wales. *Proceedings of the Geologists Association, London*, Vol. 93, 91–111.

HORTON, A. and HORRELL, J. 1971. Field meeting in the Peterborough district. Report by the Directors. *Proceedings of the Geologists Association, London*, Vol. 82, 353–358.

— LAKE, R. D., BISSON, G. and COPPACK, B. C. 1974. The geology of Peterborough. *Report of the Institute of Geological Sciences*, No. 73/12, 86 pp.

KENNARD, A. S. and WOODWARD, B. B. 1922. The post-Pliocene non-marine mollusca of the east of England. *Proceedings of the Geologists Association, London*, Vol. 33, 104 pp.

KENT, P. E. 1947. A deep boring at North Creake, Norfolk. *Geological Magazine*, No. **84**, 2–18.

— 1962. A borehole to basement rocks at Glinton, near Peterborough. *Proceedings of the Geological Society, London*, No. 1595, 40–42.

MACFADYEN, W. A. 1933. The foraminifera of the Fenland clays at St. Germans near King's Lynn. *Geological Magazine*, Vol. 70, 182–191.

— 1938. Post-glacial foraminifera from the English Fenlands. *Geological Magazine*, Vol. 75, 409–417.

— 1970. The microfauna of four samples of post-glacial silt and clay from the Coronation Channel, Spalding. Appendix II. 161–164 *in* The stratigraphy of the northern Fenland. SMITH, A. G. *in* The Fenland in Roman times. PHILLIPS, C. W. (editor). *Research Series of the Royal Geographical Society of London*, No. 5. 360 pp.

MARR, J. E. and KING, W. B. R. 1928. Pleistocene Marine Gravels at Eye, Northamptonshire. *Geological Magazine*, Vol. 65, 210–212.

MITCHELL, G. F., PENNY, L. F., SHOTTON, F. W. and WEST, R. G. 1973. A correlation of Quaternary deposits in the British Isles. *Special Report of the Geological Society, London*, No. 4, 99 pp.

MORGAN, A. 1969. A Pleistocene fauna and flora from Great Billing, Northamptonshire, England. *Opuscula Entomologica*, Vol. 34, 109–129.

PORTER, H. 1861. *The geology of Peterborough and its vicinity*. 126 pp. (Peterborough: private publication.)

PRENTICE, J. E. 1950. The sub-surface geology of the Lincolnshire Fenland. *Transactions of the Lincolnshire Naturalists Union*, Vol. 12, 136–137.

SABINE, P. A. 1949. The source of some erratics from north-eastern Northamptonshire and Huntingdonshire. *Geological Magazine*, Vol. 86, 255–260.

SALWAY, P. 1970. The Roman Fenland. 1–21 *in* The Fenland in Roman Times. PHILLIPS, C. W. (editor). *Research Series of the Royal Geographical Society, London*, No. 5. 357 pp.

SHENNAN, I. 1982. Interpretation of Flandrian sea-level data from the Fenland, England. *Proceedings of the Geologists Association, London*, Vol. 93, 53–62.

SKERTCHLY, S. B. J. 1877. The geology of the Fenland. *Memoir of the Geological Survey of Great Britain*. 335 pp.

SMITH, N. J. P., JACKSON, D. I., ARMSTRONG, E. J., MULHOLLAND, P., JONES, S., AULD, H., BULAT, J., SWALLOW, J. L., QUINN, M. F., OATES, N. K. and BENNETT, J. R. P. 1985. Map 1. Pre-Permian geology of the United Kingdom (South). British Geological Survey, Keyworth, Nottingham.

— (compiler). 1985. Map 2. Contours on top of the Pre-Permian surface of the United Kingdom (South). British Geological Survey, Keyworth, Nottingham.

TAYLOR, J. H. 1963. Geology of the country around Kettering, Corby and Oundle. *Memoir of the Geological Survey of Great Britain*. 149 pp.

TOOLEY, M. J. 1982. Sea-level changes in northern England. *Proceedings of the Geologists Association, London*, Vol. 93, 43–52.

APPENDIX 1

List of Geological Survey photographs

Copies of these photographs are deposited for public reference in the Library of the British Geological Survey, Keyworth, Nottinghamshire. Prints are available on application. The photographs belong to Series A.

5634 Crown Pit, Fletton [c.19 96]
Mechanical excavation of Oxford Clay (1931). The face is 10 m high.

5635–36 Beevy's Pit, Fletton [c.19 96]
Mechanical excavation and 13 m face in Oxford Clay (1931).

5637–8 Beeby's L.B. Pit, Fletton [c.19 96]
Mechanical excavation of Oxford Clay. Both photographs show the removal of the topmost 4 m or so of weathered clay, the 'Callow' by mechanical belt to worked out areas of the pit where it is deposited in ridges. The underlying well stratified compact shaly clay is used in the Fletton process of brick-making.

5639 Typical Fenland Scenery, Guyhirn [c.40 04]
View across Waldersea.

11636 River Nene Floodplain. Ham Farm, Orton Waterville. [TL 1535 9734]
Section in the floodplain meadow showing crumbly textured humic alluvial clay resting on gravels of the First Terrace.

11637–638 Bishop Benedict's Tomb, St Peters Cathedral, Peterborough [TL 1943 9864]
The ornamental coping of the tomb is Alwalton Marble, a shelly limestone which is thought to be developed in the lower part of the Blisworth Clay near Alwalton Village.

11640 The Whittlesey Brick-pits [TL 1961 9366]
View north-north-eastwards from the glacial-drift ridge north-east of Yaxley. The brick-pits are built on the outcrop of the Fen peat and gravel deposits but extend down into the Lower Oxford Clay.

11641 Peterborough Cathedral [TL 2000 9457]
View northwards from the glacial-drift ridge, north of Farcett over the Fletton brickworks to Peterborough.

11642–45 Woodston Beds [TL 1737 9650]
Temporary exposure at an interchange on the Western Primary Road. Gravels of the Third Terrace rest upon oxidised silts and sands and humic silty clay of the Woodston Beds.

11646–48 First Terrace Gravel [TL 2112 9856]
Section at Storey's Bar Road, Peterborough, Cambridgeshire. The dark soil is developed from night spoil and domestic refuse. The underlying terrace deposit consists of medium-grained limestone-rich gravel with seams of sand and silt. Kellaways Beds occur in the floor of the excavation. A frost wedge is visible on 11647.

11649 Blisworth Clay. Section at the A47 interchange, near Longthorpe [TL 1567 9853]
A grassed mound of fill is built on the Cornbrash, a pale brown weathering rubbly limestone. The exposed face shows the entire Blisworth Clay sequence with the underlying Blisworth Limestone at the base.

11650 Blisworth Clay. Section at the A47 interchange, near Longthorpe [TL 1567 9853]
Dark tenaceous clays of the Blisworth Clay rest upon a reddish brown-stained slightly nodular ironstone band. The beds become calcareous and fossiliferous below.

11651–11654 Weston Primary Road Excavation, near Longthorpe [TL 1578 9444]
Section extends from the Cornbrash rubbly limestone at the surface, through the Blisworth Clay to the top of the Blisworth Limestone. The last is being excavated with difficulty.

APPENDIX 2

Borehole catalogue

All the boreholes drilled for BGS in the district are listed. These include boreholes undertaken for the Industrial Minerals Assessment Unit (IMAU) as part of an investigation of local sand and gravel resources; and shallow gouge-auger and percussion boreholes related to research on the Quaternary succession. Except for the oil company boreholes, the remaining boreholes were selected because they were completely or partly cored and have detailed descriptive lithological logs. The total number of boreholes currently registered are noted. The borehole numbers are those of the BGS 6-inch records system.

TF 10 NW	52 boreholes
6	Etton Waterworks [1424 0518]. 43.6 m. Cored, Cornbrash to Grantham Formation. 1934
28 to 48	BGS Industrial Mineral Assessment Unit boreholes. 1976
TF 10 NE	53 boreholes
1	BP Glinton No. 1 [1502 0526]. 390.4 m Oxford Clay to Basement. P E Kent. 1962. *Proceedings of the Geological Society*, No. 1595, pp.40–42.
25 to 49	BGS IMAU boreholes. 1976
50	BGS Welland Wash No. 5 Borehole [1923 0824]. 6.7 m. Flandrian sequence. 1978
TF 10 SW	14 boreholes on Peterborough Sheet
10 to 12	BGS IMAU boreholes. 1976
TF 10 SE	840 boreholes
115	Peterborough Development Corporation 11E/5936. [1759 0136]. 13.1 m. Cored, Blisworth Clay and Limestone, Upper Estuarine 'Series'. 1970
785 to 794 and 798	BGS IMAU boreholes. 1976–77
TF 11 NE	5 boreholes on Peterborough Sheet
36 and 41	BGS IMAU boreholes. 1979–80
TF 11 SW	23 boreholes on Peterborough Sheet
29 to 41	BGS IMAU boreholes. 1980
TF 11 SE	44 boreholes
15 to 40	BGS IMAU boreholes. 1979–80
44	Elm Farm, Langtoft Fen [1521 1290]. 66.4 m. Kellaways Beds to ?Upper Lias; Cored, Upper Estuarine 'Series' to Lincolnshire Limestone. 1983
TF 20 NW	42 boreholes
6 to 31	BGS IMAU boreholes. 1975–76
32	BGS Welland Wash No. 1 [2138 0950]. 7.1 m. Flandrian deposits. 1978
TF 20 NE	35 boreholes
6 to 30	BGS IMAU boreholes. 1975–76
31 to 35	BGS auger boreholes, 1.6 to 3.75 m. Flandrian deposits. 1979
TF 20 SW	193 boreholes
129 to 155	BGS IMAU boreholes. 1976–77
157 to 162	BGS Catswater Farm boreholes 1 to 6. 4.4 to 36.6 m. Oxford Clay. 1978–79
TF 20 SE	47 boreholes
3 to 28	BGS IMAU boreholes. 1975–76.
29	BGS Catswater Farm Borehole 7 [2500 0384]. 7.0 m. West Walton Beds. 1979
TF 21 NW	3 boreholes on Peterborough Sheet
17, 26 and 30	BGS IMAU boreholes. 1979
TF 21 NE	2 boreholes on Peterborough Sheet
18	BGS auger borehole, Moulton West Fen No. 1 [2859 1502]. 9.6 m. Flandrian sequence. 1980
23	BGS Welland Wash No. 6 [2563 1527]. 7.6 m. Flandrian sequence. 1980
TF 21 SW	56 boreholes
2	Texaco Spalding No. 1 [2434 1478]. 494.4 m. West Walton Beds to Basement. 1971
2	Welland and Nene River Division, Anglian Water Authority, Crowland No. 1 (New Crowland Bridge) [2298 1067]. 94.5 m. Drift on Oxford Clay to Upper Lias. Cored, Blisworth Limestone to Upper Lias. 1973
6 to 8	BGS Welland Wash boreholes 2 to 4, 10.5 to 18.8 m. Flandrian sequence. 1978
16 to 44	BGS IMAU boreholes. 1979
TF 21 SE	14 boreholes
8 to 14	BGS auger boreholes. 4.8 to 7.9 m. Flandrian sequence. 1979
TF 30 NW	5 boreholes
5	BGS auger borehole, Archers Drove Farm [3126 0663]. 4.9 m. Flandrian sequence. 1979
TF 30 NE	20 boreholes
16	BGS Plash Farm Borehole [3873 0530]. 7.3 m. Flandrian–Pleistocene sequence. 1979
18 to 20	BGS auger boreholes, 4.6 to 8.9 m. Flandrian sequence. 1979
TF 30 SW	17 boreholes
8	BGS Eldernell Borehole [3275 0047]. 15 m. Flandrian–Pleistocene sequence. 1979
TF 30 SE	48 boreholes
39 to 42	BGS auger boreholes. 3.2 to 3.8 m. Flandrian sequence. 1979
43	BGS Guyhirn Wash Borehole [3810 0198]. 21.5 m. Flandrian–Pleistocene sequence.
45 to 46	BGS auger boreholes. 7.8 m. Flandrian sequence. 1980
47 to 48	BGS piston-sampler boreholes. 10.6 m. Flandrian sequence. 1980
TF 31 SW	13 boreholes
8 to 11	BGS auger boreholes. 1.8 to 7.3 m. Flandrian sequence. 1980
12	BGS piston-sampler borehole, Sycamore Farm [3370 1113]. 9.45 m. Flandrian–Pleistocene sequence. 1979
13	BGS auger borehole [3489 1156]. 4.45 m. Flandrian sequence. 1980

TF 31 SE	26 boreholes
5A	BGS auger and piston-sampler borehole, Parson Drove Borehole [3793 1052]. 10.75 m. Flandrian – Pleistocene sequence. 1979
5B	BGS Parson Drove Borehole [3793 1052]. 149.1 m. Cored West Walton Beds to Upper Lias. 1979
6 to 7 and 9 to 14	BGS auger boreholes. 1.3 to 8.2 m. Flandrian sequence. 1980.
TF 40 NW	1 borehole on Peterborough Sheet
1	Texaco Wisbech No. 1 Borehole [4066 0842]. 322.2 m. West Walton Beds to Basement. 1971
TL 19 NW	197 boreholes on Peterborough Sheet
136 to 145	BGS IMAU boreholes. 1975
270	BGS Orton Waterville No. 1 Borehole [1496 9677]. 6.9 m. Pleistocene sequence. 1976
271	BGS Orton Waterville No. 2 Borehole [1498 9675]. 7.2 m. Pleistocene sequence. 1976
TL 19 NE	More than 780 boreholes on Peterborough Sheet
36	Mitchell Construction Company Wharf Works No. 1 Borehole [1853 9797]. 13.4 m. Cored, Cornbrash to Upper Estuarine 'Series'. 1969
218/16 and 17	Peterborough Telephone Extension boreholes 7 [1905 9840] and 8 [1905 9850]. 17.4 and 26.0 m. Cored, Cornbrash to Upper Lias. 1971
249	Peterborough Development Corporation Borehole 510 832 (Peterscourt) [1951 9883]. 30 m. Cored, Cornbrash to Upper Lias. 1970
1181 and 1182	British Sugar Corporation boreholes DR2 [1110 9766] and DR3 [1770 9763]. 21.7 and 23.2 m. Cored, Upper Estuarine 'Series' to Upper Lias. 1978
TL 29 NW	255 boreholes on Peterborough Sheet
216 to 243	BGS IMAU boreholes. 1975 – 76
244 to 245	BGS Crease Bank auger boreholes. 5.3 and 5.5 m. Flandrian sequence. 1980
246	BGS Must Farm auger borehole [2325 9657]. 6.0 m. Flandrian sequence. 1980
247	BGS Stanground auger borehole [2089 9747]. 3.5 m. Flandrian sequence. 1980
248 to 254 and 258 – 259	BGS Field's End Bridge piston-sampler (No. 1) and auger boreholes. 1.6 to 6.4 m. Flandrian sequence. 1980
TL 29 NE	47 boreholes on Peterborough Sheet
28 to 35, 37 to 39, 41 to 43 and 46 – 54	BGS IMAU boreholes. 1975 – 77
TL 39 NW	17 boreholes on Peterborough Sheet
TL 39 NE	10 boreholes on Peterborough Sheet
TL 49 NW	26 boreholes on Peterborough Sheet

INDEX

Page numbers in italics refer to figures, tables and plates.

BRITISH GEOLOGICAL SURVEY

Keyworth, Nottingham NG12 5GG
Plumtree (060 77) 6111

Murchison House, West Mains Road,
Edinburgh EH9 3LA 031-667 1000

London Information Office, Geological Museum,
Exhibition Road, London SW7 2DE
01-589 4090

The full range of Survey publications is available through the Sales Desks at Keyworth and Murchison House, Edinburgh. Selected items can be bought at the BGS London Information Office, and orders are accepted here for all publications. The adjacent Geological Museum bookshop stocks the more popular books for sale over the counter. Most BGS books and reports are listed in HMSO's Sectional List 45, and can be bought from HMSO and through HMSO agents and retailers. Maps are listed in the BGS Map Catalogue and the Ordnance Survey's Trade Catalogue, and can be bought from Ordnance Survey agents as well as from BGS.

The British Geological Survey carries out the geological survey of Great Britain and Northern Ireland (the latter as an agency service for the government of Northern Ireland), and of the surrounding continental shelf, as well as its basic research projects. It also undertakes programmes of British technical aid in geology in developing countries as arranged by the Overseas Development Administration.

The British Geological Survey is a component body of the Natural Environment Research Council.

Maps and diagrams in this book use topography based on Ordnance Survey mapping

HER MAJESTY'S STATIONERY OFFICE

HMSO publications are available from:

HMSO Publications Centre
(Mail and telephone orders)
PO Box 276, London SW8 5DT
Telephone orders 01-873 9090
General enquiries 01-873 0011
Queueing system in operation for both numbers

HMSO Bookshops
49 High Holborn, London WC1V 6HB
01-873 0011 (Counter service only)
258 Broad Street, Birmingham B1 2HE
021-643 3740
Southey House, 33 Wine Street, Bristol BS1 2BQ
(0272) 264306
9 Princess Street, Manchester M60 8AS
061-834 7201
80 Chichester Street, Belfast BT1 4JY
(0232) 238451
71 Lothian Road, Edinburgh EH3 9AZ
031-228 4181

HMSO's Accredited Agents
(see Yellow Pages)

And through good booksellers